MICHEL SERRES AND THE CRISES OF THE CONTEMPORARY

Michel Serres and Material Futures

Serres is a radically disruptive thinker with respect to the history of philosophy, the practice of philosophy and the future of enquiry. His writings stretch from the sixties of the last century to up to the minute discussions of ecological and economic crisis.

This book series will provide discussion both of the various stages and aspects of his own writings, and extend the discussion to developing responses to questions and problems of and for philosophy, which his writings have brought to the attention of his readers. The return of a concern in philosophy for mathematical innovation is one such area of interest, as indeed is the focus on ecology and climate change

OTHER TITLES IN THE SERIES:

Mathematics and Information in the Philosophy of Michel Serres, Vera Bühlmann

MICHEL SERRES AND THE CRISES OF THE CONTEMPORARY

Edited By
RICK DOLPHIJN

BLOOMSBURY ACADEMIC
LONDON • NEW YORK • OXFORD • NEW DELHI • SYDNEY

BLOOMSBURY ACADEMIC
Bloomsbury Publishing Plc
50 Bedford Square, London, WC1B 3DP, UK
1385 Broadway, New York, NY 10018, USA

BLOOMSBURY, BLOOMSBURY ACADEMIC and the Diana logo are trademarks of
Bloomsbury Publishing Plc

First published in Great Britain 2019

Cover design: Irene Martinez-Costa

A catalogue record for this book is available from the British Library.

A catalog record for this book is available from the Library of Congress.

ISBN: HB: 978-1-3500-6069-2
ePDF: 978-1-3500-6070-8
eBook: 978-1-3500-6071-5

Series: Michel Serres and Material Futures

Typeset by Newgen KnowledgeWorks Pvt. Ltd., Chennai, India
Printed and bound in Great Britain

To find out more about our authors and books visit www.bloomsbury.com
and sign up for our newsletters.

CONTENTS

SERIES EDITORS' PREFACE

Everything communicates with everything else. This simple principle has guided Serres' meticulous and inventive examination of the material world, and the sense to be made of it, since the publication of his earliest writings over fifty years ago. His peregrinations have been directed by his readings of German polymath Georg Wilhelm Leibniz, and of Roman atomist Lucretius, and informed by responsiveness to current developments in mathematical topology and information theory. The juxtaposition of old and new, careful reading of historical text and sensitivity to current context are distinctive.

The immense range of his writing is marked by a profound respect for the specificity of the matter in hand. Serres moves cautiously from the local to the global. His writing navigates a route not just through a problematic, or within a given discipline, but also across the disciplines, between the sciences, philosophy, literature, mathematics, mythology, art, history and anthropology, and from problem domain to problem domain.

Serres' work is experimental, dynamic and open. At a time when more of his writing is becoming available in English translation, the aims of this new series of publications are to develop a positive and critical reception of this work and to create opportunities for those inspired by it to continue to innovate.

Joanna Hodge
David Webb

ABOUT THE AUTHORS

Eugenie Brinkema is associate professor at the Massachusetts Institute of Technology. Her articles have appeared in numerous journals including *Angelaki, Camera Obscura, Criticism, differences, Journal of Speculative Philosophy, qui parle* and *World Picture*. Her first book, *The Forms of the Affects*, was published by Duke University Press in 2014.

Vera Bühlmann is professor for architectural theory at the Technical University in Vienna, and head of its Department of Architectural Theory and Philosophy of Technics. Together with Ludger Hovestadt, she is also head of the applied virtuality lab at the Institute for Technology in Architecture ITA at ETH Zurich. She is the author of *Mathematics and Information in the Philosophy of Michel Serres*, which appears in this same book series.

Rick Dolphijn is associate professor at Utrecht University and honorary associate professor at Hong Kong University (2017–2020). He wrote *Foodscapes* (2014), *New Materialism* (2012) with Iris van der Tuin, and edited *This Deleuzian Century* (2015) and *Philosophy After Nature* (2017), both with Rosi Braidotti. He is now finishing work on 'the Cracks of the Contemporary'.

Gray Kochhar-Lindgren is professor and director of the Common Core at the University of Hong Kong. Previously, he served as associate vice chancellor at the University of Washington-Bothell, as a Fulbright Scholar, and as a visiting professor at Utrecht University's ICON. Currently, he is writing on transdisciplinarity, philosophy and the city.

Patricia MacCormack is Professor of Continental Philosophy at Anglia Ruskin University. She is the author of *Cinesexuality* (2008) and *Posthuman Ethics* (2012), the editor of *The Animal Catalyst: Toward Ahuman Theory*

(2014) and the co-editor of *Deleuze and the Schizoanalysis of Cinema* (2008), *Deleuze and the Animal* (2017) and *Ecosophical Aesthetics* (2018). She publishes extensively in the posthuman, queer theory, animal studies, horror film and Continental Philosophy.

Lucie Kim-Chi Mercier is lecturer in philosophy at the Centre for Research in Modern European Philosophy (CRMEP), Kingston University, London, where she obtained her PhD in 2015. Her research focuses on the philosophy of translation, epistemology and history, 1960s French thought, as well as postcolonial and critical race philosophy. She is a member of the Radical Philosophy editorial collective.

Matteo Pasquinelli is professor of media philosophy at the University of Arts and Design, Karlsruhe, where he is coordinating the research group on critical machine intelligence, KIM. He recently edited the anthology *Alleys of Your Mind: Augmented Intelligence and Its Traumas* (Meson Press), among other books. For Verso Books he is preparing a monograph provisionally titled 'The Eye of the Master: Capital as Computation Cognition'. Website: http://matteopasquinelli.org.

Janina Pigaht is a documentary filmmaker and impact producer. She has developed and produced three documentary films, of which her first won the debut competition at the Netherlands Film Festival in 2013. As an impact producer she works with artists to make them see how their work can contribute to change.

Joeri Visser is a PhD candidate at the Institute for Cultural Inquiry from the University of Utrecht. His research focuses on Deleuze and Guattari's biopolitics, the language of a minor literature and the interrelated commitment to immanence combined with a close reading of poems, letters and essays by Artaud. As a translator of Serres, he has published a translated lecture in Philosophy After Nature in 2017. Since 2011, he has also been teaching French language and literature at a secondary school near Rotterdam.

David Webb is professor of philosophy at Staffordshire University. He has published work on Michel Serres, Michel Foucault, Gaston Bachelard and Jean Cavaillès. His published work on Michel Serres includes 'Penser le multiple sans le concept: vers un intellect démocratique' in *Michel Serres*, edited by F. Yvonnet, 87–94 (Paris: Éditions de l'Herne, 2010); and 'Michel Serres on Lucretius: atomism, science and ethics', *Angelaki* 11.2 (2006), 125–136. He has co-translated a new edition of Michel Serres' *The Birth of Physics* for Rowman & Littlefield (2018).

1

Introduction: Michel Serres and the times

Rick Dolphijn

Over the past decades, interest in the work of Michel Serres, a longtime member of the *Académie Française* and renowned philosopher since the late 1960s, has undergone a renaissance. Without abandoning the philosophers dear to him (Gottfried Wilhelm Leibniz, Blaise Pascal, Lucretius, to name a few), his lyrical prose or the wide range of academic and non-academic perspectives that has characterized his 'generalist' style from the beginning, Serres' writings today capture the urgencies of the times more convincingly than anyone else's; from the digital revolution to the ecological crisis to the future of the university, *all* the crises that mark the world today are addressed in a lucid, affirmative and remarkably original analysis by Serres. Publishing an average of one book a year (reaching up to four volumes in 2015 alone), his immensely creative and diverse perspectives on the 'times of crisis', or 'the contemporary' as we will call it from now on, have already given him the status of a philosophy superstar in his home country. There, *all* of his books are now becoming bestsellers, especially since the publication of *Petite poucette* in 2007, which sold over 300,000 copies in France alone. In addition, Serres has a radio show in France (with Michel Polacco) which has up to four million listeners weekly. I do not know of any philosopher, alive or dead, who has been able to attract such audiences, or whose ideas are so thoroughly grounded in our various intellectual traditions, yet so entangled with the here and now. Following the massive sales of *Petite poucette* (translated into English by Daniel Smith as *Thumbelina*) in France, Serres is gaining more and more readership in the Anglophone world. With a dozen volumes currently translated into English (there are over seventy titles available in French, so we are still talking about only a fraction of his oeuvre) and now a book series with Bloomsbury Press in which his philosophy is being explored, serious attempts are being undertaken to establish Serres, well into his eighties now, as a leading twenty-first-century philosopher. Finally.

A philosophy of the contemporary

This volume aims to explore the way Michel Serres thinks the contemporary, the way this has always been key to his thinking and the way this has intensified over the past decades. To do this, the contributors to the volume aim not so much to write 'about' Serres ('explaining' his thoughts) as to write 'with' him. In a way, their aim is to 'work' with his writings; to explore how they resonate with the world. This comes very close to the way Serres himself puts the philosophers, artists and scientists that interest him in step with the events and the experiences that he has lived through. From his first publication on Leibniz (*Le Système de Leibniz et ses modèles mathématiques*, published in 1968), his breakthrough, it was clear that Serres was not impressed with the disciplinary (i.e. arbitrary and constraining) borders that are so important to academia today (more so than ever), but which in the end – of course – aim only at safeguarding a field of study, or the positions of power it secures. Being occupied with Leibniz while occupying Leibniz, Serres never limited his analyses to the reception of Leibniz in philosophy, to the themes dominant in philosophy or even to the structures of twentieth-century academia. For Serres, the book is not opposed to the world, it has no subject (no particular idea functions as its essence) and it has no object (it has no strict relationship with its outside world). This means that rather than taking 'a book' as his point of departure, Serres, much like a proper premodern philosopher, begins by picking up the many different resonances which 'occur' as he reads the book and the world, or better, from which both the book and the world come to be. Starting from these resonances means that the philosopher is to analyse how the pink world paints the Panther Pink, while the Pink Panther paints the world pink (to paraphrase Gilles Deleuze and Félix Guattari). Rather than having a subject or an object, philosophy is about the reading/writing of the concept (pink in this case) and empowering all the senses, and the aim of philosophy is then to map this concept across all of its appearances (in all that it *necessarily* resonates with). This is the original task of philosophy; to become aware of, *and* to realize, a wholly new world.

In order to train one's soul as a philosopher, one must wrap oneself around this world, and search for ways to live it. This is why Michel Serres always lived an adventurous life. Born in Aquitaine in the south of France in 1930, he was raised on the banks of the Garonne river, far removed from the intelligentsia in Paris. After naval academy, where he majored in mathematics and physics, and some years of travelling the seas of the world, the currents of the times carried him to further his study in philosophy and the humanities. Starting his academic career in Clermont-Ferrand, where he shared an office with Michel Foucault, he was asked by Foucault to come to Vincennes, at the margins of Paris, where he found his new home (and where the interview, later in this book, also took place). Nevertheless, Serres has spent much of his life travelling, teaching at many different universities around the globe. And notably, he always combines this with hiking, rock climbing and of course travelling the seven seas. This is a philosopher living a life, whose philosophy is about thinking a life, whose ideas are inextricable from the river he grew up by, and from the exhilaration of climbing icy rocks.

The deep green and the bright blue have to be introduced into philosophy. For Serres, just as there is no world outside the book, there is no life 'outside' of philosophy. The tea houses in Rome matter to the tragedies of the Manhattan Project, as the thoughts of Empedocles matter to the ecological disasters taking place at the poles of our planet. Serres had to be there when the seas were dying (and he saw how we murdered them!). He experienced it through his own body. How could he not live (and die) through this, without rethinking Homer and Conrad? On the seas, Serres heard the wind cry for help, which once again showed that one never cries alone. Our companions are all around us (and not just in the 'species' that we can distinguish). We have only to search for ways to recognize them. The first task of philosophy is then to recognize such resonances, to feel them. Or, as Serres (2013, 37) puts it: 'I am initially what pain has made of my body; after that, long after, I am what I think' (*own translation*). This is what philosophy looks like, immediately moving away (in one simple and perfect stroke) from Cartesian dualism, the Cartesian Cogito and the Cartesian disregard of Nature, by offering us a wholly other alternative.

In one of his more recent books, Serres analyses the financial crisis, but he begins by saying something very important; namely, that the word 'crisis', coming from the Greek *krino*, actually means 'to judge'. He calls for the setting up of a new institution or tribunal in which the here and now, both visible (the stock market crisis) and invisible (the deep structures of capitalism), can be re-evaluated. Serres immediately adds that 'the crisis of all crises', being the ecological crisis, is in the end more central to our times than capitalism and modernism (which can best be regarded as the two horses that speed the ecological carriage . . . as expected, in the direction of a cliff). Reading the invisible, or deep structures that accompany the present, it doesn't take Serres very long to end up with René Descartes. Descartes (Serres 2012, 28), giving us the idea of the modern age, proposed that we become like 'the masters and possessors of nature'. Of course, this idea reveals itself in many different forms in the ecological crisis as we witness it today. It encapsulates the dualism that lies at the heart of our complete and tragic misunderstanding of 'nature' (see also Serres [2017]). If the times so clearly ask us to set up an institution to evaluate all the crises that require judgement today, we *have* to take into account the many ideas, artworks and books that accompany our times, that cannot be left out of the equation when we map the crises in all of their complexity. (Of course, good ideas, artworks and books are necessarily timely, *id est*, they somehow always matter.)

This is what Serres sees as the task of philosophy: 'Philosophy is an anticipation of future thoughts and practices' (Serres and Latour 1990, 86). It has no power, like the sciences, but *in the name of the future* it has to speak up. It has to make use of its unrestrainable ability to compose 'a world, *in its totality, or in general, and in its most minute detail*' (168; emphasis in the original). And it has to offer this world, not just to academia, but to everyone. This book proposes thinking the contemporary with Michel Serres, to give birth to worlds of the here and now to which we have become blind.

To feel all the resonances, or to open up our minds to everything that happens with the times, the contributors to this volume, analyse various crises of the contemporary by situating Serres' institutions in the different crises of the contemporary. Serres wrote a great deal about how such institutions might be established in such a way as to give proper attention to the ecological crisis. Thinking the earth anew, Serres (2012, 31) distilled a concept and set up a new institution: 'This is what I call "Biogea", an archaic and new country, inert and alive, water, air, fire, the earth, the flora and fauna and all the living species.' This most fundamental new institution he called 'WAFEL', which stands for water, air, fire, earth and life. It is an institution which, through the sciences, the arts and philosophy (the different modes of thought available to us), could be given a voice that takes into account the earth as a whole, without being dominated by any elite (religious, capitalist or modernist).

The need for new institutions is a call for *transdisciplinary thinking* and for being occupied with the present as it occupies us, as I put it elsewhere (see Dolphijn 2012). In an older text, Serres (1982, 58) gives a timely example of why we need to open up all aspects of the present to our many different traditions in thought as it is only then that we can see the emergence of a new space, a new time. 'No one can draw the edge of a cloud, the borderline of the aleatory where particles waver and melt, at least to our eyes. There a new time is being fired in the oven. On these totally new edges, which geometry and the art of drawing have abandoned, a new world will soon discover dissolution, atomic and molecular dissemination.' Let us call on mythology, topology and new media art to teach us about these new times! We will never know what a body can do, but it is our responsibility to the earth, and to ourselves, to care, and thus to always question our presuppositions, to reach out to the unknown of the body, to the matters that matter.

A crisis is what matters

Throughout the book, important concepts that Michel Serres has been working with for decades (think of the parasite, *'amis de viellesse'*, softness and hardness) serve as vital tools for analysis. These concepts are put to work in the movies we see, they appear in the buildings we live in, they traverse the institutions at which we teach and learn. To think the contemporary with the concepts that Serres has offered us, and not (implicitly or explicitly) with the Cartesian Dualist epistemologies that have obscured our vision, allows the contributors, careful readers of Serres from many different parts of academia, to *write the world as it occurs*, showing us this world to which we have been blind for so long. By asking the questions that only philosophy can ask, which means that we anticipate the knowledges and practices to come (as we read in the epigraph to this book), philosophy's task now is to *raise the world that matters*, as a whole.

Raising the world that matters is impossible through critique. In an interview with Bruno Latour, Serres once said that '[a]n idea opposed to another idea is always the same idea, albeit affected by the negative sign. The more you oppose one another, the more you remain in the same framework of thought' (Serres and Latour 1990, 81). The dominance of Cartesian philosophy, modernist thinking or dualism cannot be shaken off by critiquing them. Rather, this is done by revealing the world that these traditions are hiding from us. Instead of commenting on the dualist others (Others), the affirmative approach that Serres takes inspires us to reveal the multiplicities, the noise, the flows of information that have made up our world from the start.

Serres aims to establish a materialist philosophy that could be considered a mathematical materialism (in the French tradition, à la Pascal) but that might better be called (after Leibniz) an algorithmic materialism, since it begins with the way information is emitted, received, stocked and processed (see also Serres in Braidotti and Dolphijn [2017, 13]). Starting with the algorithm and the way it traverses the organic and the inorganic, the mind and the body, nature and culture, the material and the immaterial, Serres neither accepts nor critiques the dualism of subject and object. His thinking traverses both (the subject and the object).

To be more precise (as this is an important issue), Serres (like Deleuze) *does not believe* in objects. The emergence of the object (which is to say our belief in them), he has stated many times (see, for instance, Serres [1995, 87]) should be considered the only identifiable difference between animal societies and our own. The object is invented to *stabilize* our relationships, to slow down the time of our revolutions. Isn't that exactly why first the church and then the state was so keen on creating objects, totems, places of worship, prisons and courthouses, making our histories slow? And isn't late capitalism, so focused on speeding up the processes of production, most of all known for having a fetish for just about every object imaginable? Serres is right when he concludes that there is no sacred without the sacred object, no war without an army and no exchange without value. Making us believe in the object seems to be necessary for the operation of power in human societies. It surely explains that the late capitalist world we live in today, in contrast to the idea that it is characterized by speed, has by now come to an almost complete stop; immobile as a traffic jam. Capitalism and modernism are forcing us to look in the stone eyes of Medusa. A whole system of objectified belief is being created in order to keep the monetary machines running. The machines that benefit those in power.

Thinking with love

The petrifying power of capitalism and modernism (among others) have given us the petrifying forces of critique (as discussed above). To avoid looking into the eyes of Medusa, the contributions in this book are unwilling to comment upon the writings of Descartes, his subject position, his dualism, his view of nature. We walk on. Being true to life, to the forces of life, to the sensations that guide life and the pains that warn of death, this book focuses with Serres on the madness and the *sophrosyne* of that which we love, that which occupies us affirmatively, that which we care for and for which we take responsibility. In *Biogea*, Serres asks us to follow Empedocles, who thought in terms of love and hate, just as Spinoza, two millennia later, thought in terms of joy and sorrow. It is the same simple lesson: let us strive for love or joy and let us move away from hate or sorrow. Longing for *Biogea* means longing for the unified voice of the world, and the hope that knowledge and wisdom will once again function together (through new institutions, transdisciplinary analysis). Serres (2012, 75–6) thinks with Empedocles when he reviews the state of academia today, suggesting the necessary changes that life asks for: 'Simple and easy, our old sciences rested on the analysis that separates and cuts up, on the cutting up that separates subjects from their objects. Hate? Difficult, global and connected, the life and Earth sciences presuppose communications, interferences, translations, distributions and passages. Love?' Not only the Garonne river but also the sea flows through Serres, resonates with all of him and thereby also flows through all of his thoughts. The sad conclusions that the sea is dying, therefore, *has* to play a role in all the ideas he proposes to us. It does not terrify him, rather, it forces him to rethink the sea anew. The only way to give birth to a new sea is to love it dearly. Thus, we have to set up the new institution to *think with* the sea, to find the allies, the resonances, that together can reveal a new sea. Biogea, where humans and the world live in symbiosis.

This counts first and foremost for the way we, in our day and age, have been alienated from the world we live that 'rethinking nature' is actually what shows our human impotence at its (Freudian) best. Let us therefore give rise to a new world, a new university and a new (posthuman/nonhuman) society. A new world, which philosophy has to compose in its finest detail. It is a new world that we love, that we occupy while it occupies us and that is offered to all as a gift from the future.

References

Dolphijn, Rick. (2012). The Revelation of a World That was Always Already There: The Creative Act as an Occupation. In: Rosi Braidotti and Rick Dolphijn, *This Deleuzian Century; Art, Activism, Life*. Leiden, Boston: Brill/Rodopi.

Serres, Michel. (1982). *Hermes: Literature, Science, Philosophy*. Baltimore and London: Johns Hopkins University Press.

Serres, Michel. (1995). *Genesis*. Trans. Geneviève James and James Nielson. Ann Arbor: University of Michigan Press.

Serres, Michel. (2012). *Biogea*. Trans. Randolph Burks. Minneapolis: Univocal.

Serres, Michel. (2013). *Variations sur le corps*. Paris: Le Pommier.

Serres, Michel. (2017). Information and Thinking. In: Rosi Braidotti and Rick Dolphijn (ed.), *Philosophy after Nature*. London and New York: Rowman and Littlefield International.

Serres, Michel (with Bruno Latour). (1990). *Conversations on Science, Culture, and Time*. Trans. Roxanne Lapidus. Ann Arbor: University of Michigan Press.

2

The virtue of sensibility

David Webb

The work of Michel Serres reveals a profound concern with the relation of human life to nature and the challenges facing technological society in the twenty-first century. Pollution and the degradation of natural resources are only a part of it. Serres reminds us that we can no longer expect the natural world to soak up the consequences of human action without discernible, and in many cases dangerous, repercussions – climate change being the most obvious and most urgent example. As society becomes coextensive with the Earth, so together they behave increasingly like a closed system in which cause determines effect and all our actions leave a lasting mark (Serres 1995, 42). Might the predictability this involves at least enhance the prospects of intervening to limit the damage caused by the reckless short-termism that characterizes much human activity, most especially in the economic sphere? Serres argues that this is not so. On the contrary, he argues that our entry into a deterministic closed system robs us of the resources we need in order to make any significant change; resources that Serres associates with the capacity for invention in making sense of the world, but also with restraint. The challenge that Serres identifies is first of all that of striking a new contract with nature in order that society and nature not be considered essentially separate domains (38). However, his proposal to revise the familiar conception of the social contract is only a part of the story. For Serres, a contract is not primarily a legal artefact at all, but an alliance or conjunction; that is, a form of coexistence born out of a material exchange (39; 2018, 151, 154). Throughout his writing Serres reflects on the material world, elaborating an ontology of flows, communication, exchange, noise and information in which order and disorder increase and decline in turn, and locally, without being subject to an overarching law. For Serres, this is fundamentally about what it means to exist, and above all what it means to be alive; as such it is a meditation on how we live, and even on how we might live well. What Serres explores, tentatively and often indirectly, is therefore less a new approach to contract theory in the usual sense than a virtue ethics; or it is both, a virtue ethics being the condition for a return to the idea of contract. Traditional forms of virtue ethics pertain to our rational faculties and the management of emotion. The conception of virtue that Serres sets out is distinctive in that it is planted in sensibility, where the energies and codes of the material world are translated into meaning and language. Such a conception of virtue may support wider cultural and political responses to the challenges presented by technological society, but it is on the idea of the virtue of sensibility itself that I will focus in this chapter.

Before turning to Serres, however, I shall outline Georges Canguilhem's conception of health, and his approach to the relation between the living being and its milieu. I refer to it here because Canguilhem's account of health opens a perspective from which the link between sensibility and virtue in Serres is particularly distinct. As such, the account of Canguilhem I present here is intended only to provide a helpful way to approach what Serres has to say on sensibility. Although I will express a reservation with the way Canguilhem understands the relation between the living being and its milieu, this should not be read as a critical reflection on what is an enormously rich body of work.

In *The Normal and the Pathological* (1989), Canguilhem confronts what had been the established view that disease and abnormality are deviations from a norm, which is taken to be fixed, and that the task of medical science is to restore the living being to that norm. Against this view, Canguilhem argued that there is no single 'normal' condition for any living being that could stand as the ideal of its health. Instead, each living being exists for the most part in a 'normal' condition that is quite precise at any given time, defined by a series of norms relating to its physiology, but continually changing. Even disease is a norm, albeit an inferior one. Most of the time, the living being exists in a dynamic state of near equilibrium, in which the norms indicating its 'steady state' are continually changing – sometimes gradually (as with age), and sometimes more rapidly (as in the case of serious illness). Normality is therefore a condition that has continually to be regained; and crucially, as the living being achieves this, it arrives at a new equilibrium rather than restoring the old. In this way, '[b]y the sole fact of its existence, the organism resolves on its own a kind of contradiction, the contradiction between stability and modification' (Canguilhem 2012, 72).

Health is characterized not by normality, but by the capacity to maintain balance in spite of being knocked around by changes to one's internal condition, the environment and the balance between them: 'What characterizes health is the possibility of transcending the norm, which defines the momentary normal, the possibility of tolerating infractions to the habitual norm and instituting new norms in new situations' (Canguilhem 1989, 196–7). To be in good health, adds Canguilhem, 'means being able to fall sick and recover, it is a biological luxury' (198–9). One might think that the capacity of an individual to do this or not is simply a matter of objective fact, and to some extent this is indeed the case, but Canguilhem insists that to be in good health is to feel in good health, to feel 'more than normal', and as such capable of creating and following new norms of life (200). This capacity is what Canguilhem calls 'normativity', and it involves the relation a living being has with its milieu.

Canguilhem explains that the concept of the 'milieu' first arose in Newtonian mechanics, and was appropriated by French mechanistic scientists to name the medium between two bodies by virtue of which an effect is propagated from one to another. This interpretation of 'milieu' later settled into the idea of what Canguilhem (2008, 103) calls 'a pure system of relations' that lent itself to scientific analysis. In this way, 'the milieu becomes a universal instrument for the dissolution of individualised organic syntheses into the anonymity of universal elements and movements' (103). Only later, when considering a body in isolation, did it begin to signify a condition that 'surrounds' a given body, and at the same time began to take on a more absolute sense. The concept was introduced into biology by Jean-Baptiste Lamarck, who maintained that the milieu acts not directly on the organism, but on its needs, and that the living being may then react to the provocation of the milieu by modifying its needs, or its means of fulfilling them (191). This idea remains central to Canguilhem's understanding of the milieu and its relation to the living body, although he is critical of the way that for Lamarck life exists in a milieu that is entirely indifferent towards it. Canguilhem traces further episodes in the history of the concept of the milieu before referring to the work of Jakob Von Uexküll and Kurt Goldstein, who present an alternative to what had, in spite of Lamarck and in agreement with the physical sciences, become the prevailing view that the milieu exerts an influence directly on the living being. In their view, and Canguilhem's, it is, on the contrary, 'the living that makes its milieu for itself' (111).

Von Uexküll establishes a basis for this view with the concept of *Umwelt*, which Canguilhem (2008, 111) describes as 'an ensemble of excitations, which have the value and signification of signals'. But not all excitations are signals. For Von Uexküll, before an excitation in the *Umwelt* can act on a living being, it must first be noticed, and this presupposes the orientation and interest of the living being. So while the physical milieu supplies any number of excitations, the living being only retains a select few. The *Umwelt* is therefore 'a milieu centered in relation to the subject of vital forces in which the living essentially consists'. In this way, the relation between the living and the milieu 'establishes itself as a debate (*Auseinandersetzung*), to which the living brings its own proper norms of appreciating situations, both dominating the milieu and accommodating itself to it' (113). The living being is treated here as an individual within a milieu, and as irreducible to the system of relations that make up that milieu. The milieu offers the living being certain ways to meet its needs, and the living being negotiates over the terms of this offer. It would be easy to assume that a healthy being will always try to dominate this negotiation in order to ensure that its terms are met, but this is not necessarily the case. Canguilhem emphasizes that 'a life which affirms itself against the milieu is a life already threatened' (113). By contrast, '[a] healthy life, a life confident in its existence, in its values, is a life of flexion, suppleness, almost softness' (113). The play between domination and accommodation may therefore be a subtle one. However, it nonetheless implies a binary view in which the living being is either winning or losing. I shall argue that Serres undoes this opposition and describes the living being, and human beings in particular, as existing in a more intimate and less conflictual relation to the material world.

For Canguilhem, the milieu proper to the living being is the world as it is perceived. Referring to the human being in particular, Canguilhem (2008, 118) describes how action is oriented and regulated by values that 'pick out quality-bearing objects and situate them in relation to each other and to him. Thus the environment to which he is supposed to react is originally centered on him and by him'. The living being perceives and orders things according to the values immanent in the interests and needs that provide an orientation towards the mass of available excitations, and these values guide the creation of the milieu. For example, I need shelter, so I perceive the world as a series of potential places to hide; or I need to drink, so the world is perceived in terms of watering holes, the distance between them, their relative safety. This is the selection by virtue of which the living being pays attention to some stimuli more than others. Canguilhem goes on to conclude the essay 'The Living and Its Milieu' with the following words:

> From this stems the insufficiency of any biology that, in complete submission to the spirit of the physico-chemical sciences, would seek to eliminate all consideration of sense from its domain. From the biological and psychological point of view, a sense is an appreciation of values in relation to a need. And for the one who experiences and lives it, a need is an irreducible, and thereby absolute, system of reference. (120)

For Canguilhem, needs are neither reducible to a physical system, nor signs of a more fundamental condition or reality (as, e.g., they would be for Friedrich Nietzsche, for whom a need has to be deciphered to reveal the configuration of the will to power that speaks through it). A need is simply what it is and should not be subjected to an interpretation that does not take it seriously. It is the principle by which the living being orients itself, and in so doing creates sense and value. Yet although the need of a living being is an absolute, it is a negotiable absolute. For needs can be modified, or prioritized, and they can be met in various ways. Needs themselves, although irreducible, exist only in relation to the milieu, which provides the healthy body with ways to meet them that it can exploit, if it is able. However, the specificity of the needs of a living being means that it filters out or ignores most of the stimuli or excitations to which it is in fact exposed, selecting only those that help it to meet its needs (or are a threat). On this reading, the Earth is a source, not merely of nourishment and shelter, but, insofar as the living being picks out signals that enable it to meet its needs, also of sense and value. Sense and value are conditioned by the needs of the living being and the ways it devises to meet them. In turn, needs are a communication between the living being and the material world.

Canguilhem's focus on the health of the body leads him to adopt a perspective that regards limitations in the living being's ability to meet its needs as arising from two factors: first, a change to the state of the living being itself (accident, illness, ageing) that demands innovation in order for the being to remain healthy; second, a condition of the milieu that demands something of the living being, or constrains it in some significant way. Such limits are local and negotiable; for example, it is generally possible to arrange the milieu differently, to move to a place where conditions are more favourable, or simply to revise the need from which the sense and value that shape the milieu initially spring. What Canguilhem does not consider is a situation in which the limits are non-negotiable; that is, where the scope for modifying an existing milieu is radically reduced, or where moving to a new location makes no difference.

In books such as *The Natural Contract*, *Rameaux*, *Biogea* and *Times of Crisis*, Serres considers our experience of meeting just these kinds of global limits. The expansion of contemporary society across the Earth has led to a situation in which the local is no longer embedded within wider global limits. Technological change reaches all corners of the world and its effects bear on all life in return. Contemporary society can extend no further and 'has nothing left in reserve' (Serres 1995, 41). The scale of the Earth in comparison with human society is no longer so great that it can absorb the effects of our actions. As a consequence, the management of our collective needs in relation to the Earth is stressed close to breaking point. When all of nature is in some way part of the human condition, the consequences of our actions ripple further than we imagine. The effects of what we do are no longer felt 'elsewhere', they impinge on everyone and this strangely unexpected recursion carries a terrible threat. Like Canguilhem, Serres writes that 'strength has reserves at its disposal' (41), by which, like Canguilhem, he means scope for innovation and ingenuity in meeting needs. Such innovation may require courage, imagination, intelligence, an ability to communicate and organize, and other qualities besides. But it will also require being able to see alternatives to the existing configuration of needs and the way we meet them; that is, the communication, however mediated, between our bodies and the material world must be changeable. But when society extends to the limits of the Earth and every recess is saturated with meaning and language (Serres 2008, 112) the reserves are filled and there is no place to which it is possible to withdraw in order, in Canguilhem's terms, to create new sense in a new or remodelled milieu. When a global system becomes a single unit, it becomes weak and loses its capacity to adapt: 'A full and rigid totality . . . can break from rigor or hardness' (Serres 1995, 41). The image exactly matches that of the unhealthy body described by Canguilhem, one unable to adapt and establish a new equilibrium. Now, for Serres, here in this sentence, the body in question is human society as a whole. But, like Plato, he entertains an analogy between society and the individual, and immediately adds: 'The more plural an individual becomes the better he lives: the same is true for societies, or for being in general' (41). In fact, this is less a statement of liberal principles than a reflection of Serres' view of the body and its role in the way we make sense of the material world.

Drawing on the analogy of sailors aboard a ship, Serres describes how in such conditions the actions of each affect all, and there is nowhere to which one can retreat, no reserve. Any disorder among the crew risks shipwreck, and everyone's life depends on preserving what may be a fragile peace. When Plato appeals to the analogy of a ship to underline the necessity of good governance, it is through the command of reason exercised by the philosopher king that order is maintained (Plato 488a-489b). Serres (1995), no stranger to the sea himself, draws a different conclusion. There is, he writes, a single unwritten law for those aboard ship, 'a nonaggression pact among seagoers, who are at the mercy of their fragility' (40). The external threat from the elements ensures that peace is maintained within the limits of the ship itself by what Serres calls 'the divine courtesy that defines the sailor' (40). As there is no way to escape the continual proximity of others, with all the potential for discord it brings, the unwritten law of courtesy lends a little suppleness to help keep the peace: one leaves room for others. Serres likens this to the natural contract he is proposing to humanity as a whole (40). Exercised collectively, such courtesy would begin to keep the local and the global from coinciding, and at least begin to reveal resources where none were apparent. In this way, it would preserve a degree of the plurality necessary not just for survival, but for what Canguilhem calls health, and Serres associates with living well.

However, for Serres the conditions of living well are not simply equivalent to those of health for Canguilhem. Opposition to the reduction of the living being to a system of impersonal forces and relations leads Canguilhem to defend a form of vitalism, but Serres does not make this move. For him, too, the living being is irreducible to such a system of relations, but in the end so is the rest of the material world. What Canguilhem resists above all is the universal application of mechanism and the determinism that comes with it. But for Serres mechanism is less a distinct paradigm than a special case within a more diverse dynamic landscape. The law governed mechanism that gives rise to determinism is the physics of solids and of closed systems, which are for Serres (2018, 91), as an atomist, special cases of a more general physics of fluids and open systems. So when society fills the Earth each human action has a precise effect on the Earth, and each effect bears in turn on human society and on the rest of life. Society is no longer a local form within a complex system on a different scale. With no reserves, the relation between society and the material world approaches the point where it is locked into a deterministic process of cause and effect. Mechanism appears as a limit case and imposes itself as the only viable way to understand our relation to the material world. Of course, although mechanism and the physics of closed systems have been dominant since at least the seventeenth century, human beings have continued to act as if the Earth were an infinite resource that could endlessly absorb the effects of our actions, and that would always be so much greater than society that it could be treated as an open system. Such a view is no longer possible, and the convergence of the local and the global extends the unforgiving law of cause and effect from the material world considered in isolation to the relation between the material world and society. Life begins to eliminate its own conditions. It is this fate that the law of courtesy applied aboard ship is intended to avert. But the analogy of the ship allows Serres to explore a further alternative as well, one represented by the helmsman.

In spite of the complexity of the weather and the sea all around, the crew of the ship, bound by the limits of the hull, operate in a rigorously closed system. By contrast, the helmsman is in the open and actively engages the surrounding elements. His will 'acts on the vessel, which acts on the obstacle, which acts on his will, in a series of circular interactions' (Serres 1995, 42). He governs through feedback loops, responding to contingent circumstances in a complex dynamic environment.[1] The expertise that this requires was once regarded as distinct from the ability to govern well and to live well with others, a view that received its clearest expression in Aristotle's distinction between the intellectual virtues *techne* and *phronesis* (Aristotle 11401a-14b30). *Techne*, skill, underpinned instrumental or productive acts carried out in circumstances that one can more or less predict and control. *Phronesis*, practical wisdom, was the ability to act well in circumstances (ethical and political) that are contingent, complex and often unpredictable. However, the distinction rests on a belief in nature as a fixed order (Aristotelian teleology or Newtonian mechanics) that we no longer take for granted. The natural world, we now realize, is a supremely complex dynamic process, and in this respect less distinct from social order than previously thought. Society is not a local and isolated phenomenon (Serres 1995, 43–4). As a consequence, politicians, he writes, can no longer afford to belong in the city alone as experts in purely human affairs, while leaving an understanding of nature to others. To confront the challenges facing us today, we all, and especially those who govern, need like the helmsman to be aware that our acts elicit a direct reaction from the Earth. Failure to do so will have a twofold effect: first, it will reinforce the distinction whereby society could in principle be governed well without regard for its relation to the material world; second, it will without our realizing it bind us still more firmly to the paradigm of mechanism that is in fact merely a limiting case, and by doing so would conceal the very reserves needed first of all to survive and then to live well.

According to Serres, avoiding this calls for the kind of virtuosity required to act in a situation where the slightest mistake brings ruin, even death. Alongside that of the helmsman, he gives the examples of the mountaineer and the musician, which suggest an expertise, a *techne*. To chart a path up the face of a mountain requires skill, as does making music. Yet the emphasis Serres places on the contingency of circumstances, and the unpredictable nature of the consequences that follow from our actions, suggests this is not simply the skill of the craftsman in the controlled environment of the workshop. In the same way as human society is not set apart from the material world, one can no longer distinguish sharply between the virtues of instrumental action and those of ethics: the distinction between *techne* and *phronesis* begins to break down. How, then, do the examples of the mountaineer and the musician sit alongside that of the helmsman using knowledge, experience, wit and sense to hold a course in conditions that change moment by moment? How does the virtuosity Serres attributes to the mountaineer and the musician extend beyond the narrow expertise of a skill? To answer this question requires a consideration of sense, and in particular of the relation between sense and the living being. And this consideration leads to a conception of virtue linked to sensibility that involves a profound connection between human life and the material world.

As a Leibnizian, Serres regards thinking as a distillation of sensibility and sensibility as permeated by the multiplicity from which thinking arises; in the idiom of information theory, sensibility is close to the noise that precedes and surrounds the sense conveyed in language. And as an atomist, Serres treats the emergence of material order as the formation of a code; that is, atoms do not combine according to fixed pre-existing laws, but create code that shapes their combinations as they go along.[2] Sense is an emergent feature of the material world to which we are exposed, even though we remain oblivious to much of the information exchanged by things and the matter within them (Serres 2018, 148; 2003, 60–1). In *Biogea*, Serres (2012, 42) writes, 'The Biogea rustles [*bruit*]; it shouts on the nether side of our languages; without them; below them; outside them; beneath these lines, before the meaning of what I am saying gushes forth.' And in *The Five Senses* (2008) he explores this priority of sense to language at length, tracing the contours of the noise and multiplicity that language translates into a settled order.

How, Serres asks, can we open ourselves to the given without speaking it? The world 'is filled with propositions' and 'no space is left unoccupied' (Serres 2008, 112). It is an old question, which Serres couches in terms of energy and information, the hard and the soft. Our bodies inhabit in the world of the hard, of energies that sting our eyes and tear our skin, whereas language is soft: we are not blown off our feet by words (113). The bridge between the hard forces that are preconceptual and the softness of their expression in language is sensibility. Across this bridge the hard and the soft form a 'mingled given' in which sense coexists with its excess, the noise from which sense has yet to emerge. In fact, Serres states explicitly that our bodies perform this mingling and stand midway (115), exposed to the energies of the material world and translating them into sense. Such an idea is not in itself original, and is found today, for example, in cognitive science and empirical psychology. But while Serres acknowledges that the translation cannot be understood without science, and therefore without language, he also reminds us that the body itself lives and survives by virtue of it. Simply to describe the translation in scientific or philosophical terms is to avoid taking part in it. This might not matter, were it not that the challenge to society Serres describes has arisen from the saturation of the Earth by sense: we see, hear, touch, smell and taste only what has already been converted into language, images, information. The transition from one to the other is not simple, and there may be no single turning point, no well-defined threshold. In the pages that follow, Serres considers how the energies of the hard and the sense of the soft mingle. The world is a mélange of order and noise from which sense is made, but there are no hard and fast rules for how this is to happen, for how the world is to be granted its own soft voice. One has to listen, not so much beyond language as through it. 'We must write as close as possible to this moving, dense proliferation, to the full capacity of the senses which is opened up in this place and given to use by the sensible. By this I mean the sensible that is designated by the infinite capacity of sense' (118). If sensibility is the passage from hard to soft, from the noise of the material world to the sense enshrined in language, Serres proposes that we edge back through language from well-defined sense towards the noise and confusion that lies before it: an anamnesis that returns not to unity but to multiplicity, that does not leave sensibility behind so much as make its home there. A fascination with intermediary states and spaces has characterized Serres' work from the beginning and there is always, for Serres, a value attached to overcoming the risk involved in occupying such a state or space. As in Serres' example of the swimmer in midstream, an equal distance from either shore of the river, it is always a question of balance, of holding one's own against forces that threaten to carry one away (Serres 2000, 4–5). More generally, it is a question of how order is sustained in an open system; that is, where equilibrium is dynamic, not static – homeorhesis, not homeostasis (Serres 2018, 77). Perfect equilibrium is impossible, and there is always at least some element out of line, however minimal, as a result of which a system sets off on one course as opposed to another (61; 1968b, 203). The tiniest change in the state of my organs requires my body to re-establish an equilibrium as best it can, as the energy received causes a change in its configuration or a new exchange of energy (Serres 1968b, 203). In the case of sense, the equilibrium is between the extremes of chaotic noise and an impoverished idiom in which the same thing is said repeatedly in the same way: a dogmatism of high walls and closed borders.

It is above all from Leibniz that Serres takes his lead, and in particular from Leibniz's account of tiny insensible perceptions (*petites perceptions*). In the 'Preface to the New Essays', Leibniz (1989, 295) describes how 'at every moment there is an infinity of perceptions in us, but without apperception and without reflection'. In a well-known example, Leibniz notes that when we hear the roar of the wave crashing on the shore, we in fact hear countless minute perceptions that we cannot sense individually. We are unconscious of them in isolation, as they are not strong enough to register, yet when grouped together we hear them. In the same way, Serres, too, treats sense as emerging through the integration of tiny perceptions. *The Five Senses* is a sustained plea for us partially to reverse this integration, to differentiate and to lower the threshold at which we become conscious of the tiny perceptions that flood our senses. In this book as perhaps nowhere else, Serres urges us to be more attentive, to expose ourselves to a world that is consistently richer and more complex than the sense we habitually make of it. Quite apart from the pleasure to be gained by a heightened awareness of these perceptions that otherwise rise and fall beneath the threshold of our attention, such attentiveness is for Serres a condition for uncovering the reserves we need in order to make sense of the world in new ways, and in doing so to establish a new contract with nature.

To make his point, Serres contrasts the figure of Orpheus to that of Ulysses (usually called Odysseus in English language versions). Homer tells the story of Ulysses resisting the deadly appeal of the Sirens by stopping the ears of his crew with wax and having them tie him to the mast. If, as Serres (2008, 125) suggests, the Sirens sing the given and the everyday world, Ulysses survives being drawn to his death by nullifying the senses. In this way, by other means, he anticipates the strategy that Leibniz would later employ as the pre-established harmony between monads.

> Whether rowing or manoeuvring the vessel, the ship's sailor-engines do not tremble with desire, nor with fear, but carry out their orders, come what may, since they cannot hear them countermanded. Before reaching the vile straits, God-Ulysses has already dictated everything that will follow to his

> monad-sailors. Thus the helmsmen on our ships blindly follow the course dictated to them, not the route they can see before them; language, not the given; the orders given, not the world they perceive. (125)

The crew, cut off from their senses, rely on language alone, following the orders given in advance: 'The Siren-given vanishes' (125). Noise is eliminated. But with it all communication is also lost. Even language is silenced and consigned to the past; a meaning that must be remembered, but which cannot be elaborated or discussed: a command.[3] In the story of the Argonauts, Orpheus, too, confronts the Sirens, but his approach is very different (Apollonius Rhodius 2008, Book IV, 885–921). He survives the pull of the Sirens by confronting the danger head on and resolving the chaotic noise into music. If Ulysses aims at the minimal, Orpheus aims at the maximal. He understands that language needs music and that music needs noise, which is 'its essential condition' (Serres 2008, 123). Perfect order without noise is redundancy and with redundancy nothing is communicated. Yet if noise is the condition of communication, it is also a threat. Successful communication requires balance, and this cannot be without a risk. As Serres writes, 'Whoever speaks is also singing beneath the words spoken, is beating out rhythm beneath the song, is diving into the background noise underneath the rhythm' (120).

Serres links this account of the Sirens to a retelling of the story of Orpheus and Eurydice. Death, he writes, will transform each of us from a vibrant speaking body into a name, an epitaph: 'Death turns us into words, words turn us into dead people' (Serres 2008, 131). When Eurydice dies from a snake bite, the passage to the Underworld is then likened to the entombment of her body in language. As he descends into the underworld to retrieve her, Orpheus too leaves the hardness of the physical world behind in favour of the softness of words. But to move in this direction is easy, Serres writes: 'merely head downhill, follow on the heels of the entropic processes of dying, towards disorder and fragmentation; go from things to representations; name, describe, reduce a thing to a set of words and phrases' (133). Philosophers are said to be particularly adept at this (130). What is difficult is to move in the opposite direction, to resist the urge to rush from sensibility to meaning, and to climb 'the vertical path towards life, creation, or incarnation' (133). This is what Orpheus attempts, as he sings Eurydice's name to bring her back to life. She begins to follow him, finding her voice and becoming flesh and blood again, but not even Orpheus can achieve this and Eurydice 'collapses back into her own shadow' (133).

The figure of Orpheus represents the link between the noise of the world and what we make of it. He is sensibility itself, between the material world and the sense enshrined in language. He is therefore all of us. Yet each of the stories Serres tells about him is the inverse of the other: confronting the Sirens, Orpheus moves from the noise of the world towards language, and in the story of Eurydice he moves from language towards noise. Thus Orpheus twice escapes a form of death by being creative. In the story of the Sirens, he avoids the threat of dispersion in the noise of the world by creating within sensibility what Serres calls a transcendental condition shared by music and language, one that reverberates through our conversations, our bodies, our society (Serres 2008, 123) (having escaped this dispersion, Orpheus is later dismembered by a gathering of Thracian women). One may simply call this 'order', but here, as elsewhere, Serres emphasizes its temporality: in making music, one establishes a rhythm, a tempo, a time. This is more than just a convenient metaphor for Serres. There can be no meaning without order, and for Serres time, rhythm, is order that has not settled into redundancy (Serres 1998, 117). To open language to music is therefore to keep sense-making close to its condition of possibility, where this condition itself is created and recreated by attending to the noise of the material world.[4] In the story of Eurydice, it is this attention that keeps Orpheus from a second death, this time in language.[5] And in spite of his failure to bring Eurydice back to life, Orpheus is the example that Serres would have us follow.

To return now to the helmsman, the mountaineer and the musician, their virtuosity consists in sensing more acutely and thereby having more options at their fingertips for how to achieve a more fully developed sense (the course, the sequence of moves or the path, the interpretation). The point is not to develop a more sophisticated interpretation of the data given, but actually to extend the scope of sensibility itself so that more is given before the intellectual exercise of interpretation begins. It is a proposal that depends on Serres' (2008, 80) thoroughgoing materialism, which is reflected in his dissatisfaction with the concept of the milieu: 'I do not wish to call the place in which I live a medium (*milieu*), I prefer to say that things mingle with each other and that I am no exception to that, I mix with the world which mixes with me.' This mixing, through the skin, the eye, the ear, is too varied to be reducible to the play between domination and accommodation that characterized the relation (the *Auseinandersetzung*) of the living being to its milieu in Canguilhem. For Serres, this relation is a continual process of exchange; more or less ordered, more or less present to consciousness.

Like Serres, Canguilhem rejects mechanism as a basis for understanding life. However, his affirmation of a vitalism that exists apart from or somehow alongside the mechanistic universe assumes precisely the possibility of separating the two. When the living being as Canguilhem describes it selects certain stimuli and orders them according to its needs to create its own milieu, it hollows out its own vital niche in what is otherwise a deterministic world. Serres, too, understands the importance of disturbing the convergence of the local and the global both to preserve and to create room for life, but for him this does not lead to vitalism. Instead, he proposes a materialism in which mechanism itself is simply a limiting case, and then he moves to keep that limiting case from being imposed. It is perhaps because he sees the vitalism of the living being as at odds with the mechanism of the natural world in general, that Canguilhem describes the living being as in effect winning or losing the battle to define its own milieu.

Serres calls on us, through attention, to attune our senses and in effect to let them convey what our language cannot yet adequately express. In this way, we may discover recesses within the world that have not already been colonized by meaning. The convergence of society and the material world might then be checked; in place of a closed system, a multiplicity of local forms of order may re-emerge, each in communication with others.[6] Without such multiplicity there is no life, writes Serres, and we owe life 'to all the gaps left by the other living things, the Earth, the atmosphere, the waters, and the flames that, in return, owe their existence to the marginal reserves that we leave them' (Serres 2000, 119). The virtue of sensibility therefore lies in participating in the emergence of order from the noise of the world without deciding in advance the form that order will take, and in doing so to exercise restraint. Only then can we leave room for new sense and value to emerge through the communication of one thing with another, and our communication with them. This is a condition for the new contract that Serres proposes we forge with nature, indeed it is in effect the contract itself. Such restraint is not a turning in on oneself, but rather an openness to the noise of the material world. And in the end, for Serres, the virtue of sensibility lies in being able to stand there, between noise and language, on the cusp of sense in the making.

It is worth noting that for Serres such a cultivation of sensibility does not entail a rejection of technology in favour of a life in some sense 'closer' to nature. He does not advocate a return to a better time. Rather, looking ahead, he encourages the cultivation of new possibilities as human life continues to evolve in and through its intimate relation to technology (Serres 2001). Sensibility is a matter not just of the body, but of the augmented body. However, there is still work to be done to determine how technology that enhances our sensibility as intended by Serres differs from that which too often reinforces the coincidence of the local and the global, and the exhaustion of the reserves on which life depends.

References

Apollonius Rhodius. (2008). *Argonautica*. Project Guttenberg. Available at: http://www.gutenberg.org/files/830/830-h/830-h.htm (accessed 24 June 2016).

Aristotle. (1982). *Nicomachean Ethics*. London: William Heinemann.

Canguilhem, Georges. (1989). *The Normal and the Pathological*. Trans. unknown. New York: Zone Books.

Canguilhem, Georges. (2008). *Knowledge of Life*. Trans. Stefanos Geroulanos and Daniela Ginsburg. New York: University of Fordham Press.

Canguilhem, Georges. (2012). *Writings on Medicine*. Trans. Stefanos Geroulanos and Todd Meyers. New York: University of Fordham Press.

Leibniz, Gottfried Wilhelm. (1989). Preface to the New Essays. In: G. Leibniz, *Philosophical Essays*. Indianapolis and Cambridge: Hackett.

Plato. (1992). *Republic*. Indianapolis and Cambridge: Hackett.

Serres, Michel. (1968a). *Hermès I: la Communication*. Paris: Éditions de Minuit.

Serres, Michel. (1968b). *Le système de Leibniz et ses modèles mathématiques*. Paris: Presses Universitaires de France.

Serres, Michel. (1995). *The Natural Contract*. Trans. Elizabeth MacArthur and William Paulson. Ann Arbor: The University of Michigan Press.

Serres, Michel. (1998). *Genesis*. Trans. Geneviève James and James Nielson. Ann Arbor: University of Michigan Press.

Serres, Michel. (2000). *The Troubadour of Knowledge*. Trans. Sheila Faria Glaser and William Paulson. Ann Arbor: The University of Michigan Press.

Serres, Michel. (2001). *Hominescence*. Paris: Éditions Le Pommier.

Serres, Michel. (2003). *L'Incandescent*. Paris: Éditions Le Pommier.

Serres, Michel. (2004). *Rameaux*. Paris: Éditions Le Pommier.

Serres, Michel. (2008). *The Five Senses*. Trans. Margaret Sankey and Peter Cowley. London: Continuum Press.

Serres, Michel. (2009). *Récits d'humanisme*. Paris: Éditions Le Pommier.

Serres, Michel. (2012). *Biogea*. Trans. Randolph Burks. Minneapolis: Univocal.

Serres, Michel. (2014). *Times of Crisis*. Trans. Anne-Marie Feenberg-Dibon. New York and London: Bloomsbury.

Serres, Michel. (2018). *The Birth of Physics*. Trans. David Webb and William Ross. London and New York: Rowman and Littlefield.

Unger, Roberto Mangabeira and Smolin, Lee. (2015). *The Singular Universe and the Reality of Time*. Cambridge: Cambridge University Press.

3

Cosmoliteracy and anthropography

Vera Bühlmann

A-cosmic philosophies have only language or politics, writing or logic. But we act physically, Michel Serres maintains in his 1990 book *The Natural Contract*.[1] He thereby launches a direct attack on the Enlightenment and Post-enlightenment philosophy, which in all its diverse guises appears to assume that cosmology, seeking to comprehend cosmological nature, must be constrained by a speculative logic, and hence is bound to remain uncritical. The rationality of such a logic cannot be embedded within the kind of general order of knowledge that modern philosophy seeks. As Immanuel Kant put it: the systematic study of cosmology seems destined to produce antinomies.[2] *Cosmology* then cannot be the ambition of critical philosophy, only *cosmopolitics* – a politics that considers a cosmos (an overall order) that is rooted in an anthropological 'nature'. Today, this gesture of philosophical anthropocentrism is met with increasing suspicion, while we can at the same time observe a renewed interest in celebrating speculative thought in a manner that seeks to liberate rationality from anthropological or historicist straitjackets imposed on it by a *Principle of Reason* that claims to be entirely functional to (deduced from, servant of) a general, global 'telos'.

Michel Serres' book has a genuine contribution to make to this emerging interest in speculative materialism/realism. For him *The Principle of Reason* describes not an ideal order that can serve as a *'natural' frame of reference* on which to base a politics that extends to the level of the global needs, but *a 'natural' contract*. A contract 'which embodies both reason and judgement' (Serres 1995, 90). He maintains that modern philosophy has not been able to consider a global nature; that for it, nature has always been local whereas the collective lives only in global history. But history, Serres maintains, remains blind to nature (7). All it knows are subjective wars, dialogical combat. Serres begins his book with a discussion of Goya's painting where two fighting men do not realize that they are both being swallowed up by quicksand. Subjective wars and dialogical combat cannot deal with the new form of violence all of humanity is beginning to experience in phenomena that indicate climate change, and the possible extinction of animal and plant species, a form of violence which Serres calls 'objective' (10). Dialectical history has tried to invade the tribunal site where Being is distributed, he maintains, but the combating parties have thereby changed position so often, over the course of time, that the predicative theory of the ontological square has turned into a historical force itself: the two diagonals across which the dialectical positions run back and forth, ceaselessly exchanging places, have thereby accelerated as they pivot around the vertical axis: battling over how, in the name of nature, things are to be defined and addressed has turned from an originally juridical site, where the distribution of proper rights of things according to their kind, and general and individual nature were at stake, into an emergent form of violence that appears to strike back against the imposed logical classifications and local orders which all compete to become referential, and to extend their scale from local to global. Disturbed, dynamized and furious, the Earth responds to this historical force. It begins to tremble and threatens to swallow up the combatants together with all those who watch the spectacle and place their bets on one of the parties.

Humanity, writes Serres, has become a physical variable (17) and it is high time to begin thinking in these terms.[3]

We ceaselessly inform global Nature through our movement and energies, Serres maintains, and it in turn informs us through its global change by the same means. The exchange of information of which Serres (1995, 109) speaks is physical: 'Our technologies make up a system of cords or traits, of exchanges of power and information, which goes from the local to the global, and the Earth answers us, from the global to the local.' The exchange of information gives birth to a kind of physics, he proposes, whose order is at once objective and multiple, and which is probabilistic and complex. It is an order that is elemental and instructive, rather than elemental and predicative. It is *physics born from communication*, which is at the same time – reciprocally and without ever coming to rest in any one state of reciprocal correspondence – a *physics of communication.*

My interest in Serres' approach here is not in strengthening facticity against conceptual instrumentalizations and agoratic competition; in fact, quite the reverse. A physics of communication, if we think it with Serres,[4] can open up a path for thinking the process of hominization beyond any presumed predication of 'the human'. A physics of communication allows for a materialist approach to hominization that cannot be accounted for by either cosmology or cosmopolitics. Serres (1995, 24) sets wit and materiality into a peculiar relation that allows beauty to be addressed, in an ethical sense as a hope for peace. It is crucial that Serres' approach to beauty is not: how can it be achieved? nor: how can we recognize it and not be mistaken? Serres takes a kind of microbiological point of view: he calls the relation with the help of which we can address beauty a relation of 'equipollence' between humanity and the world, between spiritedness and materiality. Both factor in nature, as equals in force, power, effectiveness, signification or validity. Serres thereby links back to earlier ideas according to which beauty shines forth, produces a gleam that reveals something true. It is a relation which brackets out the essentialist question regarding nature from the scene of action (the scene of action in Goya's painting, considered as the ontological square). For a communicational physics, questions about who sent the messages, are they reaching their destination and what or who holds sway over a faithful transmission can be bracketed along with the search for predicative answers to the question of nature's essence. This is because a communicational physics, like the corresponding materialist view regarding hominization, *neither pursues a single determinative answer, nor neglects the questions* – it seeks to appreciate the beauty of the nature to which such a communicational physics gives birth: 'Can we practice a diligent religion of the world?' Serres asks (48). Diligence is decisive, and its opposite is negligence. Because the identity (being) of such a nature lies in its beauty and can only be sustained in communication like a secret sustained by its currency, by having it circulate without ever exposing it, by referring to it without ever wanting to determine its meaning exhaustively – this is what it means to follow his approach to communication via cryptography. Serres (2016, 168) writes: '[N]ature is hidden twice. First by the cipher. Then with an ingenuity, a modesty, a subtlety, that prevents our reading the cipher even from an open book. Nature hides beneath a hidden cipher. Experimentation and intervention consist in bringing it to light.' The nature of a communicational physics can be addressed only indirectly, in quasi-referential plays that dramatize the placeholder positions (cf. Serres 1982b, Bühlmann 2018). Because of the vicariousness of the space at stake, the terms of such a communicational physics are contractual in the sense that they are binding for all the parties involved. The obligation of the natural contract is to keep the secret of nature's beauty without mutilating it, so that it can shine forth and radiate: 'Out of the equivalence, the identity, the fusion of the worldwide world and the worldly world arises beauty. Thus it surpasses the real in the direction of the human and the human in the direction of the real, and in both cases sublimates both' (Serres 1995, 24). The nature at stake in a communicational physics can be neither possessed nor dominated. Serres reframes the central question of *The Natural Contract*, that regarding humanity having become a physical variable in the planetary ecosystem, by writing: 'To anyone who detaches himself from battles because even average wisdom makes them seem vain, if not inhuman, or who does not want to pay for his worst desires with infamy, the worldwide world today offers the painful face of mutilated beauty. Will the strange and timid radiance of dawn be harmed by our brutality?' (24).

Of course this sounds quite miraculous, *cryptic* even, but the clue (if I may say so) that keeps us firmly planted on the secular side is that Serres' communicational physics treats nature like thermodynamics treats energy: in purely quantitative but qualifiable terms, as the indefinite yet invariant magnitude that is conserved in all the transformations that happen in time and that is manifest in space – whatever this 'energy' or, as Serres has it, this 'nature' may 'essentially' be. Nature does not feature as a variable in this equation; it is humanity, as the keeper of this nature's secret beauty, that features as a variable within nature. But neither does this nature feature directly as a constant, providing rational roots, and determinate values of so-called coefficients. This is what, in classical physics, so-called natural constants are supposed to do. It was the great achievement of the mathematician Emmy Noether to have provided theoretical physics with a formulation of the natural constants in algebraic terms, as conservational laws (Noether 1918, 235–57). The very idea of nature being governed by laws has since lost its direction, entering a chance-bound and matrix-like cloud of directionality: the telenomy of natural direction must then be related to an initial, indefinite invariance, and all that science can say is how its magnitude can be conserved through all transformations. The conservation of an invariant magnitude can be treated as coextensive with the conservation of textual meaning by translation: the invariant magnitude needs to be mapped in terms of symmetry structures that can be translated into each other. Of course, translation cannot ever by achieved perfectly. Such an idea of perfection presupposes that the meaning of a text could be determined and recognized without contingency from the beginning. This in turn presupposes not only the idea of an original, pure, Adamitic language in which such unambiguous meaning could be formulated, it also seals off a domain of meaning from the reality of things that are becoming.

Serres' suggestion is so radical – neither classically materialist nor classically idealist – because it maintains that nature consists in the *form-bearing charges* that are exchanged in the communication between the two poles of a delicate and critical, because *genuinely unlikely*, relation of equipollence: *that between the Earth and Humanity*. I want to suggest that the knowledge which constitutes the diligence, or negligence, by which science articulates this one relation that matters above all others can be addressed through *cosmoliteracy*. The acuity and sensitivity with which it does this determines the qualitative richness of nature as it is conserved through earthly and human activity, and the capacity of these qualities to coexist. In his little booklet praising Ilya Prigogine's critique of the principle role of closure in thermodynamics, and by implication thermodynamics' dismissal of relations of equipollence, Serres stresses Prigogine's point: order out of fluctuation, he says, is not something new, but rather the very definition of novelty (Prigogine, Stengers and Serres 1991). If we settle with this peculiar relation, *equipollence*, we can find a manner of relating to 'modernity' such that it might find a way of continuing in a way that is consistent with its own values: namely, a disregard of authority claimed on no other grounds than those of tradition. We must leave the domain of *global history* for the domain of *global nature*, Serres urges. Scientific knowledge is knowledge which responds to its object, the Earth. But it neither possesses nor dominates it; rather it acts as an equal to it in terms of force, power and validity. Scientific knowledge and the Earth are to be considered equals in terms of effect, power and signification. It is the terms of this contractual relation that can be articulated with greater or lesser diligence, or negligence.

The sameness at stake is a sameness that rests within itself, but that never actually comes to rest: it is a sameness that is vibrantly catching up with itself, ceaselessly seeking to comprehend all that it, in its virtual actuality, encompasses.[5] Serres' notion of the Earth is a delicate one, a fragile one that draws, for all we know, on an experience of genuine unlikeness – *it is a mistake to assume that rationality and the real are most proximate in kind* (cf. Potter 2000). They are unlike and their sameness is genuinely unlikely: Serres calls the harmony of the rational and the real a miracle (Serres 1995, 24). The relation of equipollence strives infinitarily to manifest itself as a reciprocal equivalence between the two, the Rational and the Real, the Earth and Humanity, an equivalence which is never fully given in any explicit manner. They can be in discord: 'If our rational could wed the real, the real our rational, our reasoned undertakings would leave no residue' (24). And indeed, he continues to explain, whenever one side is considered to be more powerful than the other, such discord arises. One side then acts on the other in violent ways because it knows it has, if needed, supplementary resources to call on. Such discord, the breakdown of this equipollence, is called 'pollution', Serres tells us (ibid.). Garbage proliferates only in this gap between the real and the rational. We have so much pollution, so much garbage today, because reason acts violently upon the world – it is not enough that each thing have sufficient reason, reason must be given back, rendered. As Serres writes: There must be an equity of exchanges (90). The sameness that rests in genuine unlikeliness *is* only in the activity within a network of cords that strives to bond all factors within it: 'It's an equation of optimization, symmetry, and justice', as Serres puts it (89). That is why to consider reason rather than law as natural is a short circuit that conceals that reason is always founded *on* a judgement. It neglects that, as Serres points out, every judgement is preceded by a trial (21, 22). And judging is equivalent to weighing, he insists (ibid.), it operates upon the most efficient algebraic method, that of a proportional analogy, as in A:B = C:D. According to this algebraic method, the resolution of equations is possible – in increasingly diligent and complex manners, not only theoretically, but also historically. In mathematical terms: if we don't restrict the numerical domains which are allowed to count, then equations whose terms are raised to arbitrarily high powers all yield solution spaces of *n* solutions. If this is ignored, reason prevents the speculative articulation of new cosmical resolutions of what is the only equation that matters – the mutually implicative and reciprocal bonds between Earth and Humanity.

'When physics was invented', Serres tells us, 'philosophers went around saying that nature was hidden under the code of algebra's numbers and letters: that the word *code* came from law' (Serres 1995, 39). For Serres, law prevails over the rationality of science, which is why law precedes geometry and algebra. Modern philosophy's mistake may have been to institute a principle of reason that is supposed to found law, but that of the philosophers of antiquity was to insist that legal contracts depend upon language, and that we can pinpoint natural law in the logical or grammatical order of words and concepts. As a consequence of this unfortunate insistence, which subjects algebra to an order of language considered ideal and perennial, the birth of physics was delayed, Serres maintains, because no one paid attention to *how the Earth speaks to us in the terms of forces, bonds and interactions*. These code-based terms are enough, he insists, to make a contract between humanity and its partner, the Earth. Nature, to Serres, is that very contract. It is the web of exchanges of information, technical *and* not, manmade *and* not, in which the physics of global nature is born out of communicative activity. In it, we must assume, *everything counts without exception* (112ff.). The urgent question is not how to separate that which ought to count from that which doesn't. It is the real that must be accounted for by rationality – not any presumed notion of the true. This translates into the registers of a natural economy for Serres:

> Reason is founded on a judgment. But who gives what, and to whom must we render reasons? The answer leaves no doubt: to all things. If every thing has its sufficient reason, we must render that reason to the very thing, well named, that we call the given. The world, globally, and phenomena, proximate, local, or remote, are given to us; it would be an injustice, a

> disequilibrium, for us to receive this given free, without ever rendering anything in return. Equity therefore demands that we render at least as much as we receive, in other words, that we do so sufficiently. (90)

It is by treating issues of subjectivity and identity in terms of a jurisdiction that is prudent rather than foundational that Serres can develop a literally materialist view on hominization: 'The process of hominization "takes" in us', he maintains, 'the way a crystal undergoes a phase change and solidifies' (101). With the advances in mathematics, rationality acquires novel capacities and capabilities – this at least is how I can make sense of the 'phase changes' of which he speaks. Such acquisition then goes hand in hand with a commitment to render back 'more' of reason to the things of the world as they are given, now, by all the inventions that come from technique and artifice. The principle of reason, to Serres too, is that reason must be sufficient. But this sufficiency, for Serres, is not qualitative, but *quantitative*: the richer in information a thing is, the more reason must be given back to obey reason's principle, which is that of sufficiency. Serres considers such an exchange within a communicational physics as follows: 'What can we render to the world that gives us the given, the totality of the given? What can we render to the nature that gives us birth and life? The balanced answer would be: the totality of our essence, reason itself.' The process of hominization is tied to rationality paying back the real, sufficiently, in reason – 'the totality of our essence'. Hominization, therefore, depends upon how such sufficiency is practiced, and there is no master plan, voice or frame of reference that can determine this sufficiency. Serres' relation of equipollence (beauty that radiates, and shines, whenever the real and the rational respect each other as equals in terms of power, force, effect, significance) is the equivalence relation of an economy, but an economy in which there is no natural tendency towards equilibrium. It is an economy, therefore, marked with inverted signs: indeed, capital is not secondary, conceived as extracted and accumulated from the redistribution of a naturally, originally, balanced stock of value; in Serres' notion of economy (1982a, 173), capital is primary – 'the real, ultimate capital is the sun'. From a scientist's perspective, a sun indeed is a kind of 'originating principle' in the universe: a sun accretes with the occurrence of nuclear fusion. And there are millions of such 'principles', even in one galaxy alone, and each one is 'original', in the sense that each one bears within itself the secret of its own singularity: a banked account, an objective record, of the unbelievable unlikeliness of some incandescent cosmic dust occurring out of nowhere, mysteriously, in such a way as to *add up*, to *join* each other, catching fire and *fusing*, and forming active particles, polytomic and chance-bound sections, restless and radiating (rather than atomic cuts through a containing continuum). But it is not only the context of physics that is evoked by Serres' notion of the sun as the ultimate capital; the other context evoked, that of economy, is just as straightforward once one takes account of Serres' inverted view according to which capital is natural and primary, rather than secondary and a result: like financial capital, a sun too is indeterminate without being infinite. The world results from a natural economy, a communicational physics, an entropic exchange of information within which islands of negative entropy form local pockets, islands of relative stability that organize in a great variety of manners. By speaking of the sun as the ultimate capital, Serres links cosmic evolution with a cosmic economy. Reason itself, the totality of humanities' essence, is what needs to be rendered back to the world that radiates in its beauty, that gives itself away in its phenomena, in its things – this we have already seen. Now we can grasp better how the totality of such an essence, reason itself, can be rendered back: 'If I dare say so, nature gives to us in kind, and we render to her in cash, in human sign currency. The given is hard; reciprocity, soft' (Serres 1995, 90). Information is not gratuitous, it can only be obtained at a price, Serres (1974) has elaborated elsewhere. To integrate more information, the settled order of the integrating agency pays the price of putting its integrity at risk, it must affirm being shaken up and unsettled in its organization, it must make itself more vulnerable. 'Does becoming human consist of forever unbinding so as to bind elsewhere and otherwise?' (101). Is this the nature of thought? Do we cast off from our local customs to join the universal only to change cords? Serres asks.

The nature of thought, like the nature of earth, must be considered generic and universal, not individual and general. Remaining within the registers of generality establishes what Serres calls a political thanatocracy (cf. Serres 1992). Its power is based on betrayal, Serres maintains already in an early essay in the *Hermès* books. It reigns by distributing death in the name of protecting life-in-general (*bios*). Thanatocracy betrays humanity from becoming human: it administrates the stock of rational potency encoded in general forms, while decoupling these specific rational potencies from their real source: this real source is the power of abstraction that renders currencies current, and information capable of circulating. But why thanatocracy? Serres (1995, 112) maintains that it is only in relation to the reality of death – and death is always singular and cannot be generalized – that one can deal with a given situation in a manner in which 'everything counts':

> Once you cast off, everything you do can be held against you. The words of the examining magistrate resound. High place: high court. Here the causal space of cases is open, with no apologies or forgiveness. Every act counts, every word and even intention, down to the slightest detail. Like a judicial proclamation, an act accomplished here is immediately performative.

The ordinary world is forgiving, because here, the cords are not taut, they are slack.

> How to define our ordinary world? 'That doesn't count': this is the only rule or, better, the gap in its laws, the cord's braids and loops. A thousand things without importance are neither obligatory nor punished here. You do not have to pay for every detail of common life. A hundred spaces beyond

> the law let you do, say, or get through as you wish. Customarily, non-law prevails over law. The ease of our bodies comes from this elbow room. Who would complain about these degrees of freedom, this gratuitousness that makes up life itself? (Ibid.)

And yet, *it is the vivification of life by death that produces intelligence and diligence*: 'Death vivifies life, which dies from lack of death. Depart – toward nature – to be born', and a few lines earlier: 'So all my stories and the whole universe are reversed: assurance puts us to sleep, ordinary life gives itself over to death, the death in which normal stupidity, repetitive and limited, slumbers, drugged and bound – whereas the other worlds are populated with the lively and hardy. The taut' (114). If scientific terms are identified as lawful terms, rather than considered as terms that need to rest within the spectrums that attempt to grasp the improbable, the unlikely, as are dealt with in jurisprudence, then there is only negligence, no diligence. Then science imposes: this or that does not count.. The universal rights produced by modern politics in this manner are not universal, because they erect a general order, *an order where a particular rationality controls the real*, and therefore a hierarchical and dynamical order, rather one of equipollent radiating actuality.

Science produces generalizations, but the true power of science does not derive from the stocks of potency stored and encapsulated in generalizations. It derives from abstraction. Abstraction doesn't extend in dimensions, it opens up dimensionalities: It is categorial, not classificatory. Generalizations render, they map abstraction's power into temporal and spatial relations, while abstraction itself transcends time and space. It concentrates around an empty center, it considers both the negativity and the positivity of a considered vertical axis. The universal can never be represented in global terms, because it is present only in abstraction – it is not only categorial, it is '*cardial*', it transcends time and space because it *pulsates* in a natural heart that nourishes both parties of the natural contract, the Earth and Humanity.[6]

Neglecting the difference between abstraction and generalization, the 'universal properties' of modern science have produced a general order which maintains itself only (i) by producing pollution, garbage, a vile residue, a latent noise, subterranean or climatic, that begins to attack this order violently from behind its own back; and then (ii) by propagating its programs of pacification against which no one dares object, *because objective violence, for this general order, means objective guilt*. And guilt is the lever with which *an order where a particular Rationality controls the Real* betrays both the Earth and Humanity.

Allow me just briefly to point out some indexes of how we could go further in making sense of Serres' postulate that 'hominization "takes" in us the way a crystal undergoes a phase change and solidifies'. In my work I am reconsidering the role of 'writing' in the history of humanist thought within the registers of algebra (frameworks of correspondence, methods of balancing), by relating writing to cryptography. This will mean thinking of algebra as a kind of information-based alphabetization in which the characters capture not voiced sound, but radiating activity. With Serres' (2018, 168) starting point for a communicational physics, namely, that 'nature is hidden twice, beneath a cipher and beneath a dexterity', we can comprehend the 'characters' of such information-based alphabeticity as the characters of 'terms' articulated according to *algebraic forms* – algebraic forms as *contractual statements* that render *objective* how to keep a balanceable relation of equipollent reciprocity. Algebraic forms so conceived are determinative too, they do determine an objectivity; but this objectivity is that of a global violence within a physical economy of information. Algebraic forms are determinative of the price that the Rational must pay back to the Real, that Reason must pay back to the Earth, that Thinking must pay back to Global Nature. The knowledge that such writing is capable of keeping and transmitting, then, is neither prophetic nor evangelic, it simply articulates the actuality of objective violence. Where global history tries to find a global horizon, a kind of Master Integral for all that happens, global nature tries to find rationalities that can account for cases whose cause appears in-determined.

It is clear that imagining such a cosmoliteracy is a speculative endeavour, but let me try to make a case for it here. It is literacy in a kind of bonding – literacy in writing according to the terms of information-based, algebraic forms, the forms of a geometry that, like pre-theoretical geometry in Antiquity, addresses the Earth through measurement and indexing, and that, like theoretical geometry (Euclid's legacy), demarcates an objective point of reference, but a geometry that does so not only with regard to the Earth's extension in space, or an extensionality in space-time (global history), but also with regard to an extensionality of recollection – which we can call the Earth's spectrality. We could think of the algebraic forms at stake as the forms of a spectral geometry, with the help of which one can find articulated, in *the real, actual and virtual world* – the world in which 'everything counts' and where at stake is the criterion of sufficiency for reason that wants to be critical, yet needs to be instructed in how to achieve its aim, which is to contract objective violence in such a way that the partners of such contracts are recognized as equals in terms of equipollence – *objective things* as rising (French: *surgir*) from the seas of information, anadyomene, genuinely unlikely and chance-bound like Aphrodite rising from the foam.[7]

The real, actual, and virtual world within which such bondage is articulated is a world in which, from the point of view of physics, particles radiate actively and are not entirely stable; they bond and decay. Serres (1995, 108) responds to such physicality with his strong notion of the cord. The cord, he tells us, is capable of establishing three practices which regard *form* (conceptual, geometric, knowledge), *energy* (material, physical, power) and *information* (judicial, legal, complexity). Responding to this radiating activity of the world, Serres' Natural Contract is literally meant to conserve *conditions of cordiality which organize the electrostatic force of a communicational physics.* A cord is capable of (i) marking out a field and surrounding it with flexibility. This is what it means to define an object in terms that are cordial, rather than determinative. (ii) It attaches a subject to this object, as if to its knowledge or property. (iii) It informs others contractually of the situation produced by the cordial enclosure (107ff.). In this way the cord is, as Serres calls it, '*a triple tress' of information, form and energy* – the 'curly cord' is to him the very texture of the material fabric of cordiality. 'All in all', as Serres elaborates, the cord, this 'triple tress links me to forms, to things, and to others, and thus initiates me into abstraction, the world, and society' (108). He continues: 'Its channels pass information, forces, and laws. . . . In a cord can be found all the objective and collective attributes of Hermes. When flexible, it embraces topology only to describe geometrical forms once it stiffens' (ibid.). But it is material, this cord, that initiates a spectrum – 'brief little pulls, low energy levels (amplitudes) to convey information', and 'when continuously pulled taut, it transmits force and power, high energy levels' (ibid.).

It is of course a poetic gesture to describe the cord in Serres' cordiality as a triple tress; but it is also a precise name, namely, for an electromagnetic field: *information, form and energy* are needed to articulate not only a curl's lively and never properly tameable activity, *but also the alert rather than dynamic activity that results from the propagation of waves*, in which the quantum particularity of exchangeable charges are vibrantly and continuously arranged. With this, we have a starting point from which to consider Serres' (1995, 101) proposal that hominization, within conditions of cordiality, "takes" in us the way a crystal undergoes a phase change and solidifies'.

Crystallization is the process of forming a reciprocally symmetric structure from a material fluid. It is an extensively studied field because depending on local-yet-universal conditions, a single fluid *can solidify into many different possible articulations with different properties*. This ability of a solid, rather, its group of atoms, particles and electrons, to exist in more than one reciprocal body (crystal form) is called *polymorphism*. The final form of the solid is determined only abstractly by the universally valid conditions under which the fluid is solidifying *locally* – conditions such as the chemistry of the fluid, the ambient pressure, the temperature and the speed with which all these parameters are changing. Crystalline structures occur in all classes of materials, with all types of chemical bonds. A chemical bond is an attraction between atoms that allows the formation of chemical substances containing two or more atoms. The bond is caused by the electrostatic force of attraction between opposite charges.[8]

Translated into our communicational physics, the natural contract would characterize an electromagnetic field between The Earth and Humanity, as poles of opposite charges. But what can we take from this for a materialist view on hominization? Let me try to disentangle this peculiarly 'univocal analogy' proposed by Serres.

1. It treats *the human* like crystallography treats *the crystal*: as an encrypted abstraction, arcane in essence, but through scientific description also encipherable and decryptable, and of material, chemical and physical effectivity in these very operations. In crystallography, the Earth is studied in terms of crystallization, and crystallization is studied in terms of the cases that can be found by experiment and attentiveness, as Serres would put it, to 'how the Earth speaks'.[9] The actuality at stake here originates in the strangeness of the object witnessed in empirical studies for which a metrical experiment set-up is necessary but never enough: in order to notice how the Earth speaks, attentiveness and acuity are required as well. Both poles of such a relation *prosper in their powers* with which they can, together, account for the apparent richness in phenomena that concern them. It is in that same manner that we could study the Earth in its mutually reciprocal relation to Humanity. This would then be *anthropography* rather than *anthropology*. It would be to study the patterns in which social nuclei are bound together in relational forms of collectivity.

2. Just as crystallography remains entirely undecided with regard to *what energy is supposed to be* – all it needs to assume is that the amount total of energy in the universe be invariant, and likewise with regard to *what matter in itself is supposed to be* – an anthropography too must remain entirely undecided with regard to questions of essentiality. It needs to assume *what life is* as little or as much as crystallography needs to assume what energy is. And it needs to predetermine what *vitality or spiritedness* is as much or as little as the latter needs to predetermine what materiality is.
3. Crystallography works with graphical notations that can be precise or imprecise. The structural fabric of this notation, with which it speculates and experiments, is graphical only because there is scripture which its graphisms articulate. But, and this is the decisive point, *this scripture is not meant to represent anything*. It must be considered simply as *striving to keep relations reciprocal* in the abstract symmetries it articulates. There is a notion of law involved, but those laws *don't represent* nature. They are universal, but virtually so, like algebraic formula. They are actualized within the constraint of locally particular conditions that are *symbolically manifest* in the algebraic, and cryptographical, forms of contracts. Perhaps we can say that nature so conceived is universal and *genital*, while it is kindred and specious (sexed) only in the articulations of such formulaic contracts. In such a word-play, then, genderedness would apply to the contractuality, as the symbolic nature of such anthropographical articulation.

Am I, in my reading of Serres, proposing to combine alchemy with mysticism, in the name of new science – the science of communicational physics? I readily admit that this cannot easily be refuted. But then, are these not flagwords brought forward to *call to reason* – and hence, effectively, to *terminate* – attempts to formulate new forms of speculative materialism/realism? My concern has been to set out how Serres' (2006) idea of a Natural Contract, which begins by insisting that issues of climatic and environmental concerns must be addressed in the terms of law and philosophy, rather than those of an ecology or a politically expanded version of (al)biology (Biopolitics), has something important to contribute to this emerging interest in the role of speculative experimentation and their conditions of computability. So let me summarize what this core contribution seems to be: There is a 'reciprocal transformation of cause into thing and fact into law', Serres maintains. 'It describes the double situation of scientific knowledge: its arbitrary convention, as all speculative theory, and the faithful and exact objectivity that underlies every application' (Serres 1995, 22).

References

Bühlmann, Vera. (2018, forthcoming). Vicarious Architectonics, Strange Objects. In: Maria Voyatzaki (ed.), *Architectural Materialisms: Nonhuman Creativity*. Edinburgh University Press.

Crahay, Anne. (1993). *Michel Serres, la mutation du cogito. Genèse du transcendantal objectif.* Paris: de Boeck.

Grier, Michelle. (2012). Kant's Critique of Metaphysics. *The Stanford Encyclopedia of Philosophy* (Summer Edition), Edward N. Zalta (ed.): https://plato.stanford.edu/archives/sum2012/entries/kant-metaphysics/.

Irigaray, Luce. (1983). *L'Oubli de l'air – chez Martin Heidegger*. Paris: Les Éditions de Minuit.

Irigaray, Luce. (2010). *Le mystère de Marie*. Paris: Les Éditions du Crieur Public.

Noether, Emmy. (1918). Invariante Variationsprobleme. In: *Nachrichten von der Gesellschaft der Wissenschaften zu Göttingen, Mathematisch-Physikalische Klasse*, 235–57.

Potter, Michael. (2000). *Reason's Nearest Kin. Philosophies of Arithmetic from Kant to Carnap*. Oxford: Oxford University Press.

Prigogine, Ilya, Isabel Stengers and Michel Serres. (1991). *Anfänge*. Berlin: Merve.

Serres, Michel. (1968–80). *Hermès I-V*. Paris: Les Editions de Minuit.

Serres, Michel. (1974a). Trahaison: La Thanatocratie. In *Hermès III: La Traduction*. Paris: Les Editions de Minuit.

Serres, Michel. (1974b). Vie, information, deuxieme principe. In *Hermès III: La Traduction*. Paris: Les Editions de Minuit.

Serres, Michel. (1982a). *The Parasite*. Trans. Lawrence R. Schehr. Baltimore and London: The Johns Hopkins University Press.

Serres, Michel. (1982b). Theory of the Quasi-Object. In *The Parasite*, trans. Lawrence R. Schehr. Baltimore and London: The Johns Hopkins University Press, 224–51.

Serres, Michel. (1992). Verrat: Thanatokratie. In *Hermes III: Übersetzung*, trans. Michael Bischoff. Berlin: Merve Verlag.

Serres, Michel. (1995). *The Natural Contract*. Trans. Elizabeth MacArthur and William Paulson. Ann Arbor: University of Michigan Press.

Serres, Michel. (2006). Revisiting the Natural Contract. In Arthur and Marilouise Kroker (eds): http://ctheory.net/ctheory_wp/revisiting-the-natural-contract/.

Serres, Michel. (2015). *Rome. First Book of Foundations*. Trans. Randolph Burks. London: Bloomsbury.

Serres, Michel. (2018). *The Birth of Physics*. Trans. David Webb and William Ross. London and New York. Rowman and Littlefield.

4

Mathematical anamneses

Lucie Kim-Chi Mercier

> *If we awaited perfection, we would never move forward. It is everywhere like here: suppose the Greeks had waited for the complete demonstration of their axioms, for their reduction to identical, geometry would still to be done. Mathematics presupposes axiomatics, but not perfection; we can develop the consequences of the former, downstream [en aval], for example by inventing differential calculus with a lot of pragmatism and little rigour, while simultaneously, going back upstream [à l'amont] of the axioms and definitions in order to logicize them. So it is with the method of Establishments: once something is agreed upon, it is beyond dispute. Likewise, without awaiting the perfection of philosophy – in this regard 'we are in a certain infancy of the world' [nous sommes dans une certaine enfance du Monde], we are at an antepythagorician stage – we have to work on its elements and build a first alphabet of human thoughts, but we must also go to the other extremity of the positive and order the proliferation of written or spoken tongues: study Chinese, hieroglyphe, cuneiform, the 'Indo-European' languages, demonstrate their harmony. Advancing one enterprise cannot not have an effect on the process of the other. By multiplying 'samples' set beyond dispute, by pluralising the results of this molecular and regional method, we can hope, bit by bit, to cover unknown territories. Hence achievement [l'achèvement] and perfection are in the end understood as goals, as horizon, not as prior conditions. (Serres 2007, 550–1)*[1]

In more than one way, Michel Serres' philosophy can be said to evolve through processes of translation: one concept is overtaken by another, whose domain of reference overlaps with the former while leaving an irreducible remainder. Transiting between mathematics, philosophy and the history of science, Serres' concepts became increasingly 'interferential' as they moved between regional, disciplinary and formal languages (On the concept of 'interference' as a mode of transdisciplinarity, see Serres [1972]). Paradigmatic of these translational processes are the *methodic* concepts (or rather *pairs* of concepts) by which Serres (1968) set out to explore the dual constitution of Gottfried Wilhelm Leibniz's scientific method in *Le Système de Leibniz et ses modèles mathématiques*. Pairs of opposites such as 'local vs. global', 'perspective vs. ichnography', 'applied vs. axiomatic' and later 'procedural vs. declarative' have continually structured Serres' philosophical investigations.

Initially, these conceptual pairs were meant to underline the novelty of Leibniz's mathematical method, a method which propelled Leibniz ahead of the Classical Age. Whereas René Descartes' reasoning was grounded on a *tabula rasa*, requiring a full understanding of A to posit B, Leibniz proceeded by method of 'Establishments' (*Établissements*), establishing hypotheses to which he could always return. In his quest for truth and universality, Leibniz equipped himself with networks of conditionality while Descartes sought to ground knowledge in apodictic or declarative certainty. Excerpted from a section of Serres' doctoral dissertation evocatively titled 'The Universal Cycle', the above passage offers a good starting point to think about the possible projection of Leibniz's *ars inveniendi* into Serres' own philosophical method, starting from the *Hermès* series (1968–80). Such a philosophy is capable of 'doing with little' (letters or elements, details or molecular samples) while keeping 'complete' totalities (languages, systems) as its horizon. Completion and rigour are not the grounds but the receding and always *futural* limits of the investigation.[2] Following Leibniz, Serres' methodic endeavour is not prescriptive but instead seeks to penetrate the logic of invention.

More recently in his *Éloge de la philosophie en langue française* (1995), Serres reframed the notion of a local, perspectival, applied mathematics from the standpoint of new technologies. Whereas the *System of Leibniz* and his early 1960s philosophy reflected Serres' interest in cybernetics, the pair 'procedural vs. declarative' emerged from a renewed enthusiasm for computational technologies at the end of the first millennium. Procedural methods, he explains in *Éloge de la philosophie*, function through algorithms, that is, through repetitive operational rules or mechanistic procedures. Mathematics, he reiterated, is equipped with a double universality: a universality that proceeds through abstraction and purification and a constructive mathematics oriented by the singular, technical and applied. The latter relies on the repetition of elementary procedures and applications that do not need to be 'mathematical' *stricto sensu*. This dual constitution of mathematics, paramount throughout Serres' early works, points to what Marcel Hénaff (2005, 183) aptly characterized as 'a double history of reason'. Indeed Serres consistently highlighted the importance of this other form of mathematical reasoning: a method that does not discriminate between mathematics and technique, mathematical 'purity' and technical inventiveness. This is not to say that Serres sought to abolish the distinction between purity and technicity. On the contrary, Serres pointed to a more profound sense of this distinction, which he developed by engaging with the philosophy of mathematics and the epistemology of science.

This chapter will tackle this crucial distinction from the vantage point of Serres' earliest philosophy of mathematics by looking at his seminal 'Les anamnèses mathématiques' (1967) alongside other essays on the origin of geometry published across the *Hermès* volumes.[3] In these texts, Serres' critique of reason focuses on two foundational myths of mathematics: the 'origin of geometry' and the 'crisis of the foundations' at the turn of the twentieth century. Working at the convergence of the philosophy of mathematics and the history of science, as well as within the intricacies of the French epistemological tradition, Serres examined the question of mathematical origins both from a transcendental and a historical point of view. After Gaston Bachelard, Serres was indeed trying to conceptualize an impure a priori against Immanuel Kant's

(1998, 134) declaration that 'absolutely no concepts must enter into it that contain anything empirical, or that the *a priori* cognition be entirely pure'. But unlike Foucault and Bachelard, Serres' reflection developed primarily on the terrain of mathematics, and more specifically at the level of the *historicity of mathematics*. In this regard, I will show that it is crucial to include Jean Cavaillès as a third name among Serres' philosophical references. Examining the distinctiveness of Serres' epistemological critique, I will argue that the concept of *translation*, paramount throughout Serres' early works, was the operator that enabled him to explore the dual constitution of mathematics as the essence of its historicity, between technicity and purity, or translation and foundation.

Autochthonous epistemology

After Auguste Comte, French philosophy of science's main impulsion has been to move away from the epistemological naturalism that had previously dominated. Such an 'exterior epistemology' could only miss the movement of science itself (see Serres 2015: 37–44). This imperative remained one of Bachelard's fundamental legacies in defining the relationship between philosophy and the sciences. As Georges Canguilhem (1977, 108) wrote in 'What Is a Scientific Ideology?', this 'requires an installation in the content of scientific enunciations (*énoncés*) and this "installation" can only be a practice' (Canguilhem 1981, 20). Through this problem, 1960s epistemologists of science were obliquely contemplating the idea of philosophy's disappearance: either 'redundant' in accompanying the workings of science, or 'logicist' but a-historical (Serres 1968, note 1, 66). As a consequence, it was through a different treatment of history that philosophy could hope to find its way to the sciences. 'If epistemology is historical, the history of science is necessarily epistemological (see 'Introduction' by Georges Canguilhem in Lecourt [2002, 9]).' For Serres as for Canguilhem, locating epistemology within the historicity of science constituted a way of resolving the problem of the 'secondary' or 'derived' status of epistemology.

Rather than articulating a 'discourse upon another discourse', an internal epistemology was to be sought *within the effective process* of mathematics. Every science was the host of an implicit theory as it unfolded in time.

'Les anamnèses mathématiques' (1967)[4] provides us with a condensed overview of Serres' approach to these questions. This lengthy article needs to be grasped in the context of the debates around the *history of truth*, which formed the immediate environment for Serres' writings at the time. Emerging simultaneously in the works of Canguilhem, Jacques Derrida and Michel Foucault, among others, this debate consisted, broadly speaking, in the reproblematization of the status of science (and its truth) in relation to its outside, be it culture, ideology or history as its conditions of production or possibility.[5] The kernel of Serres' reflection is a problem that is formulated in 'The Mathematical Anamneses' as follows. Mathematics can be understood as a 'well-formed language'. This 'pure logos' should be, as such, impervious to historicity (containing an *invariable* truth). At the same time, it seems that its truth can only be established 'by reference to the global system that contains it and makes it possible' ('Les anamnèses mathématiques', in Serres [1968, 78]). Yet, how can mathematics be at once *autonomous* and *heteronomous*? The paradox dissolves, Serres claims, if we consider the history of mathematics as '*the (meta)morphoses of a logos referred to* itself – mathematics being the science of this auto-reference, and rigour, the science of this application' (ibid.; emphasis in the original). This enigmatic formula announces in a nutshell what I will disentangle in this chapter. In order to unfold Serres' claim I will refer to other texts on mathematics published in the *Hermès* series. I will show that Serres' original contribution to the aforementioned epistemological debates is not so much to consider science or mathematics as a language, as to consider it as a permanent translation of itself and endow it with a singular form of historicity. Indeed, the reflexive process of mathematics upon itself does not lead to an infinite abyss of pure reflection, but is marked by the cultural and historical 'impurities' stemming from the irreducibly historical character of languages. I will show that, although mathematics remains the paramount example of a self-grounding discourse and an autochthonous language, it is also a fundamentally *impure* language. For Serres, 'mathematical historicity is nothing else than the history of an impurity, which means of a certain type of non-mathematicity' (92–3). Grasping the specificity of mathematical historicity requires us to show how the latter is at once a self-grounding language and a historical one. This search for a historical definition of the mathematical a priori can therefore be held as constituting Serres' own strategy to historicize the transcendental.[6] Yet, historicizing here does not mean objectifying or naturalizing its process from without, but, importantly, adopting an inner perspective on mathematics. As I will unravel in detail, this inner perspective is none other than that provided by the process of translation itself. Indeed, Serres claims that by translating itself into new languages, by translating its 'atoms of sense' into new idioms, mathematics unceasingly transforms its own grounds.

Mathematical historicity

Serres' reflections on the self-grounding character of mathematics owe much to the philosophy of Jean Cavaillès. Like Cavaillès, Serres considered that the philosophy of mathematics could not remain unchanged after the so-called crisis in the foundations and the far-reaching questioning of the foundations of mathematical activity that this crisis had sparked off.[7] According to Cavaillès, this crisis revealed that mathematics had to be considered as an autonomous becoming, a *sui generis* historicity (Cavaillès and Lautman 1994, 595–630). For him, the crisis of mathematics opened by Gödel's incompleteness theorem[8] entailed the absence of any apodictic insurance to start with: 'one needs to entrust the canonical process, the indefinite iteration of its use. And thus the deductive sequence is essentially creative of the contents that it reaches' (Cavaillès 1960, 73). In his unfinished and posthumous *On Logic and the Theory of Science*, Cavaillès proposed to grasp the demonstrative process, that is, the essence of mathematical historicity, through two basic operations: paradigmation and thematization. *Paradigmation* designates the operations of abstraction and substitution by which a structure can be deduced from a given operation; it is fundamentally oriented towards its objects. In opposition to the longitudinal character of paradigmation, *thematization* refers to a vertical movement, a reflexive reversal of thought towards the meaning of its operations. These two perspectives on what Hourya Benis-Sinaceur (1987, 25) calls the 'motor effect of abstraction' were crucial in establishing the 'singular becoming of mathematics' (Cavaillès and Lautman 1994, 594) as autonomous, necessary and unpredictable at the same time. In Cavaillès' (1960, 24) own words, the structure of science 'displays, in its movement, the principle of its necessity. Structure speaks about itself'.[9]

From his earliest writings onwards, Serres generalized Cavaillès' proposition; he made the problem of the fundaments the main vector of mathematical transformation, putting forward the *crisis*[10] as historical principle against the presumptuousness of epistemology.[11] As he wrote in 'The Quarrel of the Ancients and the Moderns' (1963), modern mathematics has the singular intention to 'take itself as object; and, in particular, as object of its own discourse' (Serres 1968, 59). 'At each moment of great systematic reconstruction', he observes, 'mathematicians become the epistemologists of their own knowledge. This transformation is a mutation effectuated from the inside' (68). The crucial point for Serres is that as much as this reflexive discourse *closes off* mathematics to the external discourse of traditional epistemology, mathematical language also *opens* itself to an always greater number of objects because it is inhabited by a movement of purification, or increased abstraction. Mathematics is not pure, it moves towards purity. Mathematical theory is, Serres argues, 'internally open, and externally closed' (72). 'The (paradoxical) result of this closure to any other domain of knowledge is that the organon, the language thus purified, becomes universal. The movement of closure is universalising' (72–3). In the precise way of a Leibnizian Monad, 'the most *independent* language is the language of languages. The least windows, the most universal reflection' (73; emphasis in the original).

> In the mirrored avenue spoken of by Lautréamont, a route is to be followed, continuous or fragmented, of light rays. This open avenue is the history of mathematics, the *history of a language* in which words strictly respond to each other, a language infinitely translated into new but homologous languages, the history of auto-referred systems, therefore closed, referring to other systems, therefore open, but referring to other systems similarly mathematical, therefore closed . . . *The history of forms* making sense within a system is thus involuted, but sometimes, and seemingly all of a sudden, taking another sense than the autochthonous, overtaking their interior auto-reference and therefore evolving outside the system, like a pathological outgrowth, towards a new internal systematic reference, like a lost ray looking for a mirror . . . *The history of truths* is always in quest of an enclosed universe which locks them upon themselves, which gives them an existence and possibility, until the requirement of rigour makes the interior application intolerable, and shatters the lock for a larger and better enclosed reference. (79; emphases in the original)

For the early Serres, mathematics is not a principle of subsumption, but one of expression, circulation and speed. Evolving between self-referentiality and invention, mathematics unceasingly needs to expand its domain of reference and thus to translate its own grounds into new languages. Serres' reflections thus re-actualize the Leibnizian problematic of the *mathesis universalis*, at once universal language and foundation of knowledge, from the standpoint of the history of mathematics: 'The tower of Babel, indefinitely rebuilt, reconstructs itself as soon as new promotions cannot use the same language with one another, nor with the previous system' (Serres, 'Les anamnèses mathématiques', 97; 1993, 25). In this sense, mathematics does not so much constitute *a* language as it incarnates a continuous search for a language; lacking definitive foundations, mathematics remains in a permanent state of crisis (Serres, 'Les anamnèses mathématiques', 106).

As a result of this mathematical revolution, Serres (1968, 55) considers that the questions of classical epistemology have become problems of 'scientific technique', rendering the 'epistemology of science' redundant (Serres, 'La querelle des anciens et des modernes', in Serres [1968, 55]). Such an internal meta-language does not limit itself to describing the course of mathematical transformations but also takes effect on the latter: 'far from stabilising or naturalising mathematics, [internal epistemology] reconstitutes it, vivifies it, restructures it' (64). Serres takes up the mutual imbrications between mathematics and this autochthonous epistemological discourse as an occasion to explore the idea of a *philosophy of the history of science*, taking a step further what Cavaillès had explored at the level of 'mathematical experience'. On the grounds of his doctoral work on Leibniz, Serres views Leibnizian systematicity as an example of 'ideal architecture' of mathematical idealities. Encompassing interrelated and parallel developments, such complex interferential architecture, involves multiple logics of temporalization, and can in no way be subsumed under a single historical narrative. Although system-oriented, Serres' 'history of the sciences' can only exist in the plural, by splitting into a multiplicity of *models* of historicity, just as the Leibnizian 'System' can only be grasped as a combined multiplicity of disciplinary languages.[12]

In this context, Serres also lays the groundwork for an analysis of the history of science as a 'history of truth'. Insofar as the history of science can be characterized as a zone of contact (*lieu de contact*) between ideality and historicity, it is situated at the clashing point between two normative systems (Serres, 'Les anamnèses mathématiques', 85). This implies the fundamental *indetermination* of the history of sciences, 'either *I know the position of the concept and I ignore its speed* (*vitesse*), . . . *or I know its speed and I ignore its position*' (ibid.; emphasis in the original). Throughout his early works, Serres repeatedly thematized this indetermination through the figures of the 'historian', the 'scientist' and the 'epistemologist'.[13] While the historian, in the manner of a documentarist, 'blindly gathers exhaustive details on a question' and so accesses the *unconscious* of science, but not its *truth*, the scientist inventor consciously intends this truth without access to its unconscious, or to the complete 'trajectory' of this truth. The epistemologist, perceived as the mediating figure between these two, needs both to 'know' and to 'circulate' (Serres, 'L'interférence théorique: tabulation et complexité'', in Serres [1972, 40]). Every ideality possesses three historical meanings (*sens*): 'its sense of birth, *i.e.* sedimented, naturalized; the whole of its senses at each reactivation'; as well as its recurrent sense for the retroactive restructuration of the whole (i.e. its scientific truth) (Serres, 'Les anamnèses mathématiques', 84). The history of science cannot be defined as a continuous tradition anymore, but only as a 'discontinuous, cut up weft (*trame*)' (ibid.). Therefore, to picture the history of science as the continuous communication of a tradition is fundamentally partial; 'it is a history that we try to make connected (*connexe*) and continuous by filling its breaks, while the scientist-inventor chops it and makes it discontinuous' (87).

Whereas Cavaillès considered that each new moment of demonstration constituted a *dialectical* movement of mathematical experience, Serres holds that each new event of mathematical history constitutes a new *translation* of mathematics. Taking a more formalist and structural route than Cavaillès, Serres arguably moves one plane 'upwards', from experience to language (*langue*), distancing himself further from any reference to subjectivity. For Serres the self-grounding character of science is entirely contained within the structures and history of its language(s). Hence the critical importance of the concept of 'translation', which encapsulates at once a series of operations of transformations per substitution (*salva veritate*) and an act of reflexive, transversal thematization, in short a '(meta)morphose'.

Purity, technicity

Beyond mathematics, it is science itself that can be rethought through this autochthonous production of temporality. Every translation *reactivates* scientific truth in a new setting. The Platonic notion of anamnesis proves pivotal. While in Plato mathematical idealities are referred to as the necessary return to their true origin 'before the cycle of incarnations' (Salanskis 2008, 145), Serres couples anamnesis with the Bachelardian notion of 'covering-up' (*recouvrement*), which functions, in this occasion, as its dialectical counterpart. While for Bachelard it was philosophy that displaced and covered-up scientific problems (Lecourt 2002, 12), for Serres the movement of *recouvrement* is a constitutive part of the historicity of science. For the latter, any reactivation leads to a certain *forgetting*, any truth can become an obstacle, and the existence of new forms only brings about new histories of science at the price of others, transforming the latter into 'modes of nescience' (Serres, 'Les anamnèses mathématiques', 91). Importantly, the latter is never an absolute situation, for past scientific truths remain valid. 'Take the field of dead histories: Greek geometry, classical analysis . . . Dead but not false: what is this death of the true which never turns to error' (110)? This, in turn, illuminates the paradoxical situation of the history of mathematical truth announced by Serres in the beginning. For Serres, truth's 'name' can only be established through a broader set of referentials, ideology, cultural formations or language. But while the concept or mode of being of truth (i.e. the philosophy of mathematics) varies, the 'automatic essence of the true' (*l'essence automatique du vrai*) remains invariant through time: 'the true remains invariant through diachronic transformations, while the *concept of truth* changes' (ibid.; emphasis in the original).

More importantly, anamnesis or reactivation is defined as a form of translation. Mathematical invention is nothing other than a 'successful application of a region upon others, or even, an application of the system on itself'. Rather than accumulating the givens of its tradition, mathematics carries its heritage over by filtering it, by always moving towards more encompassing syntheses and greater formal purity. The history of mathematics is, as such, the history of the theory of theory, whereby 'the science of science substitutes itself for science itself' (104). The original movement of science 'defines a system of translations. Each synchronic cut possesses its conditions of translatability' (105). The purifying process of mathematics is such that, we can always translate an anterior language into a posterior one, but the inverse is not true. While Euclidian space can be translated into the language of topology, the reverse does not hold: 'the intersection between two repertories can be empty' (ibid.). The history of mathematical systems can hence be grasped as 'a tran-slation, resumed in every instant, a history of discovering and re-covering (*découvertes et recouvrements*)' (106). Yet, and this is key, 'the translating correspondence fails as soon as it succeeds', there is no perfect application (107). There is always a residual impurity that is later taken up and pursued further. This unthought of mathematical history corresponds to what has not been translated into the new language, but may come back, centuries later, once rendered intelligible through a new language. This ever-changing zone is the residue, which, produced by the passage through different translations, is not entirely captured in the new language or the new form, yet reachable from another language or another point of departure. In other words, if every translation entails a certain *recouvrement*, translation is the continuous fabrication of the unconscious of science. Serres conceives of these mathematical 'untranslatables' as the dynamic core (*moteur*) of its historicity: 'the origin of history is starting anew at each translation into a new language' (108). This clarifies our initial proposition, according to which the history of mathematics designates the transformations of a language that is at once *self-referential* and *applied*: although mathematics is (we may add 'syntactically') a truly self-referential language, it is ('semantically') fundamentally non-transparent to itself. Mathematical truth can only manifest itself historically through idioms that function as its successive materializations and the inherent imperfection of these languages constitutes the dynamic core of mathematical historicity.

A corollary of this is that mathematical 'purity' is a fundamentally *relative* notion. The movement of mathematics towards a more and more refined purity and thus towards greater 'applicability' always retrospectively reflects the anterior stage as 'more technical' (Serres, 'L'interférence théorique: tabulation et complexité', 51). Therefore, by moving towards rigour and universality, mathematics also discovers its other origins: singular, applied, technical (52). This discovery can only be made from within the process of mathematics. Indeed, 'a cultural formation is only accessible as pre-mathematical within and through the autochthonous process of mathematics' (Serres, 'Les anamnèses mathématiques', 102), and never externally to it. In this sense, mathematics constitutes, according to Serres, an 'archaeological' form of research: by evolving towards purity, mathematics deepens its empirical ground, its practical unconscious (ibid.).[14] The temporality of mathematics is fundamentally dual, oriented at the same time towards its *telos* and towards its beginnings. In other terms there is no legitimate and illegitimate 'origin' of mathematical truth as each translational invention can be conceived as one; 'any origin is the origin as such' (99). Furthermore, 'these two limits, these two "origins" (pure vs. empirical) *can only exist by means of one another*, the first being arbitrarily as technical, the second as pure as one wants' (Serres, 'L'interférence théorique: tabulation et complexité', 52; emphasis mine). From this standpoint, the 'miracle' of geometry's birth can only appear scandalous: what is miraculous in it is not purity, but the arbitrariness of this act, which 'designates as pure a mixed and complex ore' (Serres, 'Les anamnèses mathématiques', 92).

Origins of geometry

Like Husserl before him, Serres' reflections on the 'origin of geometry' extend largely beyond mathematics, providing an occasion to reflect on the origin of science and the birth of philosophy. On the one hand, the origination of mathematical idealities is held as a paradigmatic case for the understanding of ideal forms in general – their genesis, historicity and mode of being. On the other hand, the hypothetical threshold constituted by this 'discovery', marking a 'before' and 'after' which is at once historical and theoretical,[15] constitutes a vantage point on historicity as such. Such association was clearly set out by the late Husserl in 'The Crisis of European Humanity and Philosophy' (1935). At the beginning of the 1960s, Derrida, who was in the same year as Serres at the *Ecole Normale Supérieure*, had published his seminal commentary-introduction to Husserl's short essay titled 'The Origin of Geometry'. This work, for which Derrida won the Cavaillès prize in 1961, contributed to the revival of this classical debate under radically new auspices. Serres' writings on the origin of geometry, which punctuate his entire early period of writings,[16] should be read in this double connexion, as a critical response to Husserl, and in parallel to Derrida's commentary.

As Derrida argues, Husserl is not interested in the *factice* historical event standing for geometry's origin; his intent is historical-transcendental as he tries to overcome the Kantian separation between innate ideal objects and empirical history, to mediate between the interior assumption of mathematical truths (or sheer Platonic anamnesis) and its purely extrinsic conditions of apparition (Jacques Derrida, 'Introduction', in Husserl [1999, 23–4]). Husserl aims to explain how geometry, in its historical development, constantly reactivates its origins and constitutes an original form of historicity, which serves as a paradigmatic case not only for its historicity but also for idealities in general. Husserl's essay is focused on the elaboration of an original historicity of science detached from the empirical history-of-facts, which nevertheless has the latter as its condition (56 and 175). This original historicity supposes the 'always reproducible, inaugural signification' (*Erstmaligkeit*) of the first concrete and lived *act* of geometrical idealization or 'proto-foundation', which each of its subsequent reactivations reopens. The continuity of tradition that Husserl describes is not ensured by its chronological continuity but by the unity of its becoming: 'it is a *history* only because it is *one* history' (38; emphases in the original). This unity depends both on the identity of the intentional act of idealization and on the identity of the language in which it is expressed. The possibility of such universal language (or language in general) is the reciprocal condition of what Husserl calls co-humanity or the awareness of constituting *one* community, of belonging to the same world (74).

For Serres, the origin of geometry is an origin not in the sense of an absolute, or sovereign beginning, but *insofar as it opens up the process of translation*, which science would thereafter not cease to be (Serres, 'Les anamnèses mathématiques', 107). Reactivation is not understood as a repetition of the intentional act of a single, 'proto-founding' and primary idealization, but becomes, in Cavaillès' sense, self-grounding. Through the concept of translation, Serres proposes to enter the effective process of science (*le procès effectif de la science*) without recourse to any notion of consciousness, grounding its 'generating necessity' in its 'material progress' (Cavaillès 1960, 78). Whereas Husserl considered the origin of geometry as an unveiling of its rational *telos*, Serres dislocates this unilinearity by giving a multiple account of the origin.

> [T]o raise the question of the Greek beginning of geometry is precisely to ask how we moved from one language to another, from one type of writing to another, from the so-called natural language and its alphabetical notation to the rigorous and systematic language of numbers, measures, axioms and reasoning *in forma*. But our documents can only *display* these two languages, on the one hand, narratives or legends and on another, demonstrations or figures . . . And so we are stationed, facing these two parallels that never touch one another. This origin is running ahead, inaccessible, un-seizable. The problem is set out (Serres, 'Origine de la géométrie, 5', in Serres [1980, 185]; emphasis mine).

The putative origin of geometry can be figured at the (impossible) convergence of two parallel lines: that of the geometers and arithmeticians and that of legends or histories. Between these formalisms and these narratives, between 'scientists' and 'historians', the 'epistemologist' can only operate punctual translations. In spite of the disparity of arguments invoked by Serres in his different narratives of the origin of geometry, each situation displays a certain relation (*rapport*), a certain passage from one realm of forms to another, hence a translational situation. As a result, univocity is not the '*a priori* and *telos*' of equivocity anymore (see Derrida 1989, 107), but the reverse is true: equivocity makes up the surroundings, and the uneliminable milieu of univocity. Translations are passages conducting from univocity to equivocity, to univocity and so on, in an unending process. Rather than 'communicating' through their common origin (that is also their *original*), mathematical idealities evolve by restlessly transporting themselves: between spaces, graphs, channels. As Serres (1993, 214) would later sum up, '[T]he history of mathematical sciences resolves the question of origin without exhausting it. It tirelessly responds to it while delivering itself from it.'

Symmetry

In 'Origine de la géométrie 4', Serres conceives the translational origin of geometry as a relation between two types of writing: Egyptian and Greek. Geometry does not emerge in the mind of the first geometer, 'Thalès or whoever it may be', but in the field of possibilities produced by the translation between two inscriptions. While hieroglyphs are figural and diagrammatic, the Greek alphabet is literal and algebraic (Delcò 1998, 21). Serres proposes we view the origin of geometry as the result of an encounter between the Egyptian skill of representation and cartography and the convention and formalism of the Greeks (Serres, 'Origine de la géométrie, 4', in Serres [1980, 179]). Geometry thus emerges from the 'short-circuit' between forms and formalism, between the hieroglyphic signaletic of words-things and the Greek metalanguage of words-signs. Here, under the heading of the origin of geometry, Serres is also examining the origin of idealities as *signs*, at the convergence of image and discourse (182). He underscores drawings and topology as prior conditions of abstraction and rereads the foundational claims of Euclidian geometry through this lens. In Serres' philosophy, mathematical idealities are historically-sedimented forms, which are to be unwound into series of heterogeneous procedures.

As Serres argues in 'What Thalès Saw . . .', first published as part of *Hermès II, L'Interférence* (1969) and greatly extended in *The Origins of Geometry*, the invention of geometry can also be grasped as a series of ruses of translations enabling the measurement of the immeasurable. He reinterprets the famous legend of Thalès at the pyramids narrated by Plutarch and Diogenes Laërtius as the making accessible of the inaccessible through the discovery of the notion of a group of similitudes (187). According to this legend, Thalès used the sun as an invariant in the determination of a series of relations, which could only be captured through the creation of a reduced model. Thalès' most important discovery lies in the construction of a prototype capable of rendering the immeasurable – the precise height of the pyramids – tangible. From this perspective, the origin of geometry is only the application of a certain relation (*rapport*) between two forms. Knowledge comes down to a technique of replication. Such an 'archaic' geometry, Serres (1993, 207) writes,

> takes its place in the open chain of those utterances and designations, but it does not provide the key to the cipher; it does not excavate the secret articulation of knowledge and practice in which the essential element of a possible origin is located . . . it measures the problem, takes its dimensions, poses it, weighs it, demonstrates it, relates it, but never resolves it.[17]

Throughout Serres' early texts on the origin of geometry, mathematicity is not located *in* pure forms, but in these applications. The rigour of mathematics is none other than an infinite development from translation to translation. Losing its character of absolute and thereby originary determination, purity becomes the result of a previous application, the making explicit of an implicit knowledge.[18] Scientificity is thus always preceded by rigour: applications and translations constitute the irreducible technicity of scientific discourse. Furthermore, Serres does not relativize the origin only by pluralizing it, but also by *symmetrizing* it, thinking the origin as a circulating reference.[19] By thinking technicity and purity as the two *limits* of science, the 'origin' becomes a liminal passage, which can be read either from the point of view of abstraction and formalization (towards purity) or from the point of view of archaeology (towards technicity). Taken together these texts configure a strongly anti-phenomenological account, not only of the history of science, but also of the so-called birth of philosophy, whereby the hermeneutical conception of the origin as *question* (which, as an address, calls for a question in return or *Rückfrage*) (Derrida 1989, 12) is overturned by a translational model. The question-response paradigm is substituted by a differential one, where movement or change arises from the necessity to respond to the crisis caused by the contact between different languages. Philosophy and science would only be born from a variation. Hence, to the phenomenological or hermeneutical account of a *tradition of truth*, Serres opposes a translational reflection on the temporally complex, multilinear *history of truth*.

As Serres would put it again in *The Origins of Geometry* (1993), 'I do not communicate with the origin through the traditional historical channel, but through the effort of invention and foundation of mathematics itself. My regression does not follow the path of the indefinite ruling out of tradition, but the vertical path of the mathematical *ars inveniendi*: It is through the latter that I reinterpret the historical tradition.'[20] Claiming an internal point of view on mathematical foundations, Serres drew a series of translational *tableaux*, in which scientific invention is equated to a liminal passage between languages. Between auto-reference and application, autochthony and heteronomy, Serres' philosophy of the history of science is both reflexive and blind to its own processes, leaving behind it a mysterious residue. By reading his reflections on the crisis of foundations and his texts on the origins of geometry alongside one another, the crucial importance of translation comes to the fore: it constitutes

Serres' original way of drawing a path between logicist and phenomenological approaches to the philosophy of mathematics, that is, as a continued reflection on the space left vacant by Cavaillès in his testamentary *On Logic and the Theory of Science*. Going back to 'Les anamnèses mathématiques' illuminates Serres' intricate references to the philosophical debates of the 1960s, many of which faded away in his successive 'anamneses' of the seminal essay. Today, alongside Bourbaki, Thalès or Euclid, they might guide us back into the warps of his singular thought.

References

Abbas, Niran. (2005). *Mapping Michel Serres*. Ann Arbor: University of Michigan Press.

Bachelard, Gaston. (2002). *The Formation of the Scientific Mind*. Trans. Mary McAllester Jones. Manchester: Clinamen.

Balibar, Etienne. (1994). *Lieux et noms de la vérité*. La Tour d'Aigues: Editions de l'Aube.

Balibar, Etienne. (2004). The History of Truth: Alain Badiou in French Philosophy. In: Peter Hallward (ed.), *Think Again: Alain Badiou and the Future of Philosophy*. London and New York: Bloomsbury Academic.

Benis-Sinaceur, Hourya. (1987). Structure et concept dans l'épistémologie mathématique de Jean Cavaillès. *Revue d'histoire des sciences* 40.1: 5–30.

Canguilhem, Georges. (1977). Qu'est-ce qu'une idéologie scientifique? In *Idéologie et rationalité dans l'histoire des sciences de la vie: nouvelles études d'histoire et de philosophie des sciences*. Paris: Vrin.

Canguilhem, Georges. (1981). What Is a Scientific Ideology. Trans. Mike Shortland. *Radical Philosophy* 29: 20–5.

Cassou-Noguès, Pierre. (2001). *De l'expérience mathématique, Essai sur la philosophie des sciences de J. Cavaillès*. Paris: Vrin.

Cavaillès, Jean. (1960). *Sur la logique et la théorie de la science*. Paris: Presses Universitaires de France.

Cavaillès, Jean. (1994). *Oeuvre complètes de philosophie des sciences*. Paris: Editions Hermann.

Cavaillès, Jean and Albert Lautman. (1994). La pensée mathématique. In: *Oeuvres complètes de philosophie des sciences*. Paris: Hermann, 595–630.

Crahay, Anne. (1988). *Michel Serres: la mutation du cogito: genèse du transcendantal objectif*. Paris: De Boeck Supérieur.

Delcò, Alessandro. (1998). *Morphologies. A Partir Du Premier Serres*. Paris: Editions Kimé.

Derrida, Jacques. (1989). *Edmund Husserl's 'Origin of Geometry': An Introduction*. Trans. John P. Leavey Jr. Lincoln: University of Nebraska Press.

Gowers, et al. (eds). (2010). *The Princeton Companion to Mathematics*. Princeton: Princeton University Press.

Husserl, Edmund. (1999). *L'Origine de la géométrie*, 3rd edn. Paris: Presses Universitaires de France.

Kant, Immanuel. (1998). *Critique of Pure Reason*. Trans. Paul Guyer and Allen W. Wood. Cambridge and New York: Cambridge University Press, 134.

Latour, Bruno. (1999). Circulating Reference, Sampling the Soil in Boa Vista. In: *Pandora's Hope: Essays on the Reality of Science Studies*. Cambridge, MA: Harvard University Press, 24–79.

Lecourt, Dominique. (2002). *L'épistémologie historique de Gaston Bachelard*. Paris: Vrin.

Peden, Knox. (2014). *Spinoza Contra Phenomenology: French Rationalism from Cavaillès to Deleuze*. Stanford: Stanford University Press.

Reverchon, Florian. (2014). Mathématique et expérience: ontologie et humanité des mathématiques. *Interphase* no. 1: 21–2.

Salanskis, Jean-Michel. (2008). *Philosophie des mathématiques*. Paris: Vrin.

Salmon, Gildas. (2013). *Les structures de l'esprit, Lévi-Strauss et les mythes*. Paris: Presses Universitaires de France – PUF.

Serres, Michel. (1968). *Hermès I, La Communication*. Paris: Editions de Minuit.

Serres, Michel. (1972). *Hermès II, L'interférence*. Paris: Les Editions de Minuit.

Serres, Michel. (1980). *Hermès V, Le passage du Nord-Ouest*. Paris: Editions de Minuit.

Serres, Michel. (1982). *Hermès. Literature, Science, Philosophy*. Edited by J. V. Harari and D. F. Bell. Baltimore and London: The John Hopkins University Press.

Serres, Michel. (1993). *Les origines de la géomérie*. Paris: Flammarion.

Serres, Michel. (1995). *Eloge de la philosophie en langue française*. Paris: Fayard.

Serres, Michel. (2007). *Le Système de Leibniz et ses modèles mathématiques, Etoiles-Schémas-Points*. Paris: Presses Universitaires de France – PUF.

Serres, Michel. (2015). Transdisciplinarity as Relative Exteriority. Trans. Lucie Mercier, in *Theory, Culture and Society* 32.5: 37–44.

Webb, David. (2013). *Foucault's Archaeology: Science and Transformation*. Edinburgh: Edinburgh University Press.

5

Kill is kiss, words are rats

Eugenie Brinkema

[T]he words of love, as is well known, sparsely, miserably repeat their one declaration, which is always the same, always already suspected of lacking love because it declares it.

– JEAN-LUC NANCY, 'L'AMOUR EN ÉCLATS'

Rich or poor, each language always implies a deterritorialization of the mouth, the tongue, and the teeth. The mouth, tongue, and teeth find their primitive territoriality in food. In giving themselves over to the articulation of sounds, the mouth, tongue, and teeth deterritorialize. Thus, there is a certain disjunction between eating and speaking, and even more, despite all appearances, between eating and writing . . . To speak, and above all to write, is to fast.

–GILLES DELEUZE AND FELIX GUATTARI, *KAFKA: POUR UNE LITTÉRATURE MINEURE*

c'est jeûner: to abstain, go without food, orient against the taking of a meal – the attuning of oneself to the tight drawn death of starvation: a malnourished speaker.

A human group is organized with one-way relations, where one eats the other and where the second cannot benefit at all from the first . . . Man is a louse for other men. Thus man is a host for other men. The flow goes one way, never the other.

– MICHEL SERRES, *LE PARASITE*

In order to discuss madness, one has to find a language.

– MICHEL SERRES, 'GÉOMETRIE DE LA FOLIE'

For madness, read horror.
Or: for horror, read love.

What is the difference between a bite and a kiss?

Bitan, bijten, beissen, from the root **bheid*, to split, to cleave, crack or pierce, rend asunder; I *bite the bullet* (and endure the pain), I *bite my tongue* (and do not inform you); if I *bite the dust*, then I die. *Bait*: enticement to bite. *Cyssan, cussen, küssen*, without a common root, is suspected to be imitative of the sound, onomatopoeic for my lips on yours, my mouth, on yours.

This is what I'm dying to tell you –

Something is happening

Noises. And also static.

The victims of the replicating virus in Bruce McDonald's horror film *Pontypool* (2008) babble, murmur, stammer and hitch, catch on words, repeat and stutter, stumble and blather, the entire confused, each giving a failed voice to the essence of impediment, all that blocks or prevents direct expression – what Jacques Lacan calls the object that, stuck in the gullet of the signifier, hard tight bone in the throat, choking catching closing clotted, can never, not ever, go down, be swallowed. Signal disturbances; thermal turmoil. Babel (and also what is brutal).

This is because the deadly infection in the cinematic world coming undone is transmitted through figures of speech: English-spoken euphemisms, 'phrases that conflict' or 'terms of endearment'. The virus having entered language, the sign of infection is a speaker who fails to speak correctly – or, rather, the mark of violence is a commuted conversationalist whose speaking has been conscripted to the motile logic of the virus, with attendant forms of contamination, invasion, infection, parasitism. Despite conforming in some ways to the meta-structure of zombie logics – some massification's attacks against an increasingly isolated group-cum-family who will eventually have to grapple with a threat from within – one could argue that the film creates not only linguistic substitutions but affective ones, replacing the seductiveness of the undead with banality, with a terrific boredom as figures disspeak, unname, play with language in place of abject spectacles of cannibalism or visceral, aesthetically vitalized frenzies of flesh. Repudiating Georges Bataille's well-known law in his 1929 essay 'Eye' that 'extreme seductiveness is probably at the boundary of horror' (Bataille 1985, 17), the law of *Pontypool* proposes rather: extreme dullness, fatigue, even lassitude, is probably at the boundary of horror. (Violence, no longer fascinating, thereby exhausts; this does not imply that violence thereby becomes exhausted.) This is, in point of fact, not quite a zombie film at all (its director insists the victim-perpetrators be called 'conversationalists') and it eschews the dominant postmodern horror logic by which the infected universe is one in which the zombie cycle preexists, and either guides or exacerbates the local trauma of undeadening. If the film gives body and mouth to William Burroughs' (1998) assertion that language is a virus (the writer envisioned 'word dust drifted from outer space'), it also seemingly enacts a performance of Heidegger's (1971, 207) insistence in *Poetry, Language, Thought* that we do not speak language, language speaks us: 'mortals live in the speaking of language. Language speaks . . . Man speaks in that he responds to language. This responding is a hearing . . . What is important is learning to live in the speaking of language'. But the film is not as Heideggerian as it might seem.

Inversion 1: Transplacing one of the central traits of zombie films since Romero's initial formulation, the unavoidable material dilemma of stupidly being stuck in and with a body – one that hungers, needs to feed, inexorably if awkwardly moves forward in space – *Pontypool* refigures the fundamental systemic problem of the subject as being an aphasic stuck in and with language. The body's pure surplus becomes the surplus play of the signifier, as metonymic slides that signal infection produce the poetic excess of negative stylization, an ecstasy of soundplay that continually departs from meaning and sense, which is to say, a turbulent production of spoken English language that undermines grounds for (critical or diegetic) understanding, generating above all a limitation of the deployment of language *for* understanding.

Finitude: what is supplanted by the noisiness of the wordless.

Language itself, always bespoken and casually advanced, ever lives on, while these sad mere speakers, nakedly scared, are bitten and broken and bombed. The cinematic consequence of this theory is that, pushed through the extremity machine that is the horror genre, lips are orgiastically eaten off, countdowns count down to black devastation screens and human civilization ends, perhaps to be totally reimagined, but first, crucially, it just ends.

Stilled: '*When a language is so strained* that it starts to stutter, or to murmur or stammer . . . *then language in its entirety reaches the limit* that marks its outside and makes it confront silence' (Deleuze, 'Bégaya-t-il . . .' 113).

Inversion 2: The figures in this film fail to learn to live in the speaking of language – the film thereby fails to adopt the stance of a Heideggerian horror metaphysics. *Pontypool* displays only the bloodied gesture of learning to die in the speaking of language.

Lovechild of Neal Stephenson's *Snow Crash*, Orson Welles' radio production of *War of the Worlds*, and based on a novel by Tony Burgess (1998), the film follows one day in the life (and death) of Grant Mazzy as he presents his morning radio show in Pontypool, Ontario. Grant and his producer, Sydney Briar, begin to receive curious reports from the outside world. Panic terror voices are fragmentary, contextless: 'Oh God . . . Look out . . . They're crazy . . . Just biting.' The increasingly hysterical calls avow, above all, that out in the cold white dawn *something is happening*. In fact, one might entirely and correctly summarize the plot of *Pontypool* as: What is happening is that something is happening.

As the referring structure of radio interplay progresses, radio form decomposes, each passed-off call not supplementing but halting and undoing comprehension. Grant is the host; Grant is a host. Moderator, receiver, but also server (of the virus); radio mediator – handing off the privilege of the speaking voice to different callers, arriving authorities, new eyewitnesses – but also the unmediated origin and master of on-air vocality. He is the one from whom the voice emerges; radio hosts *host* speech, house it that they might send it forth. Speech is *theirs* only insofar as it is continually given away. Communication technology is rendered a constant manic referring in the service of a fantasy of representational control – that everything is speakable, that the world's complexity is containable as a singular unity in the voice of the jockey. But Grant is increasingly unsure about what exactly it is that he is hosting: he is unable to reveal more than this empty formula of the something-that-has-happened in Pontypool, and the framed presence of the message 'On Air', which signals the classical media triad of presence, immediacy and transmission, starts to, in close-ups of its red neon text, testify, rather, to how this very system of communication is itself residing on so much fragile aeriform. To host language is not necessarily to find oneself at home in language.

The groundlessness of ether is also the producer's greatest anxiety: this drive to make language continually noisily present depends on forestalling the finitude that is bonded to the parlance of ending.

First point of emphasis: What radio hosts must at all costs avoid is, of course, *dead* air.

Grant floats, then rejects, various explanations for the rioting and cannibalism reportedly taking place a few kilometres away, and we can see these meta-narratives as lures for criticism as well: Is it a problem of the terrorist? Are we dealing with separatists? If, indeed, those doing violence are neither organized nor political, then what is it precisely that has *happened*? The film disregards these possible interpretations just as surely as critics of horror wedded to allegory must set them aside. Indeed, even in the ostensibly originary moment of the virus, the question of the event's meaning is subordinated *in its posing* to brute sound. The film opens with the bare rendering of vocal wave frequencies, blue against a black screen, as a voice retroactively revealed to be Grant's reads a play of assonance:

> Mrs. French's cat is missing. The signs are posted all over town. 'Have you seen Honey?' We've all seen the posters, but nobody has seen Honey the cat. Nobody. Until last Thursday morning, when Miss Colette Piscine swerved her car to miss Honey the cat as she drove across a bridge. Well, this bridge, now slightly damaged, is a bit of a local treasure and even has its own fancy name: Pont de Flaque. Now, Collette, that sounds like *Culotte*. That's 'panty' in French. And *Piscine* means Pool. Panty pool. *Flaque* also means pool in French, so Colete Piscine, in French Panty Pool, drives over the Pont de Flaque, the Pont de Pool if you will, to avoid hitting Mrs. French's cat that has been missing in Pontypool. Pontypool. Pontypool. Panty pool. Pont de Flaque. What does it mean? Well, Norman Mailer, he had an interesting theory that he used to explain the strange coincidences in the aftermath of the JFK assassination. In the wake of huge events, after them and before them, physical details, they spasm for a moment; they sort of unlock and when they come back into focus they suddenly coincide in a weird way. Street names and birthdates and middle names, all kind of superfluous things appear related to each other. It's a ripple effect. So, what does it mean? Well . . . It means something's going to happen. Something big. But then, something's always about to happen.

At the utterance of 'sounds like', this microcosm of the viral problem opens up an expanding bubble in the vertical displacements of the sound waves, framing a graphic space for the red letters TYPO that will eventually be surrounded to announce paratextually the title of the film PONTYPOOL. The allegorical wonder with which Grant ends his monologue – *What does it mean?* – is returned, or referred, only to the blankness of its privileged formula: 'It means that something is going to happen.'

Second point of emphasis: Something is always going to happen.

(In this way, perhaps it seems to you that *Pontypool* hews to the form of the novel as provided by Gilles Deleuze and Félix Guattari: the novella, they posit, is 'organized around the question "What happened? Whatever could have happened?"' (the crime, the kiss, the crime of the kiss, etc.), while the tale, its opposite, asks, 'What is going to happen? Something is always going to happen, come to pass' (1987, 192). The novel, combining the two, 'integrates elements of the novella and the tale into the variation of its perpetual living present (*duration*)' (1987, 192) – and thus *Pontypool*'s insistence – as tale – that something is going to happen and its simultaneous novella-esque panicked reaction to that thing which has happened, the catastrophe purportedly taking place *out there*, suggests just such a durational integration and ongoing present. Except for this: this perpetual living present is only *there* in order to end. The novelistic form is offered in order for the visible present world itself, in the final black screen signalling catastrophic violence, to no longer register any dimension of time. Film form begins *so that* it issues its ending; endingness requires the absolute exclusion of something that is going to happen and yet is entirely dependent – parasitic – on what has happened as what is going to happen. This is a game played out on flesh.)

When *Pontypool* finally presents a full account of that something that is going to happen, it runs like this: There is a virus that is transmitted neither through blood nor air nor bodies; *nevertheless, it is here.* Certain words, particularly phrases that signify care, affection's locutions, are infected and spread the virus when the contaminated word is spoken; language is the host that, through the repetitions that language is, enters the world, copies itself in the understanding of speaking subjects. Before the virus is fully understood, a cryptic warning arrives in a voice that breaks into the station's signal, advising, in French, that one avoid the fraught figurations of endearment. Yet that very message ends with the notice, 'Please do not translate this message', a message that can only be received once its injunction has been violated. The warning that thus explains how repetition is the mode through which the virus replicates itself arrives in the film already doubled; the warning prohibiting translation, comprehensible to the English speakers only after the six-word interdiction has been violated, arrives having already infected the form of the film.

The viral word was never not *there*.

In an account with which both a Derridean and a historian of linguistic strife in Canada might do much, written and spoken French are safe. McDonald (qtd in Bacon, 54) temporalizes his film's structure thusly:

> There are three stages to this virus. The first stage is you might begin to repeat a word. Something gets stuck. And usually it's words that are terms of endearment like *sweetheart* or *honey*. The second stage is your language becomes scrambled and you can't express yourself properly. The third stage you become so distraught at your condition that the only way out of the situation you feel, as an infected person, is to try and chew your way through the mouth of another person.

If the ground zero of every undead apocalyptic film is the structure containment-ultimately-leaks, it is here the uncontainedness of a language to which subjects are condemned that breaks through – bodies as fields of meat are incidental.

Voices and mouths, lips and teeth, however, are not. Something in the voice will continually resist this figuration of the something that is happening here. Not its *grain* (as Roland Barthes gives the voice its body) but its *gauge*, the voice's manner of capaciously measuring the magnitude of its potential referrals. Call it the measuring mouth of the voice. This is not reducible to the visual representation of speaking (extreme close-ups of the maw or those cyan sound waves) nor to an inverted metaphysics that would privilege writing over speaking. Horror is not a content but a structure; put another way, horror here is formalized, which is to say it is withdrawn, quarantined from violence.

Possibility: But all this might have something to do with a kiss.

Kill is kiss

Grant and Sydney are the last two protagonists standing. Upon leaving the shelter of the sound booth, they encounter an infected girl who lunges at Sydney, biting her neck; she and Grant beat the child to death. Taking refuge in an enclosed closet, Sydney begins to write on the wall, 'My name is Sydney Briar and today I killed a girl. I am so sorry. I am a good person.' Over an indeterminate temporal ellipsis, the walls are revealed to be covered with Sydney's confessional script, giving presence to a highly repetitive and often confused announcement of her amplifying guilt. As though literalizing Barthes' (1977, 143) insistence that to write is 'to reach that point where only language acts, "performs", and not "me"', in the temporally curious pause Sydney's writing arrives at the point where only that acts. This script, quasi-subjective, energized and forceful, performs as the repeated inscription of ethical horror. In her increasing understanding of what transpired on the body of the brutalized conversationalist – the specificity that names the one-way relation that was her force against the other – Sydney starts to fixate on the ethically non-neutral word 'kill', sign of infection, consequence of the bite. She falls to the hitch, the hiccup, the anguished tic that announces her sickness, intoning *kill, kill, kill; kill, kill, kill*. Voicing the poetic charge of the Russian Formalists, Grant, confronting her decomposing state, wonders: 'How do you take a word and make it strange?' Against the backdrop of Sydney's muttering, he settles on the determination: 'You kill the word that's killing you.'

Thus, to save the sick woman, Grant implores over and over again, '"Kill" isn't "kill"; "kill" is loving; "kill" is baby; "kill" is mayonnaise . . . "kill" is kiss'. His solution for disinfection takes the form of a language game, reminiscent of the Oulipo S+7 procedure, a method that replaces every noun in a piece of text with the noun found seven places away in some chosen dictionary. The seeking of new literary structures becomes equivalent to the work of remaining alive;

constraint engenders an antidote to blind (here: infected) repetition. This strategy works; from Sydney's intonation of the new form, '"kill" is kiss', she does not succumb to infection, and at her plea, finally, for Grant to 'kill me', his lips plant on hers, soft, the newly solidified definition, granting health and temporary closure of the signifying chain.

Inversion 3: Kiss mouth substitutes for gnawing mouth.

An olfactory ethics, via a consideration of 'perineal distance, perineal vicinity' as what is 'so private that it is mine, maximally', runs according to Michel Serres (2007, 145): 'You shall love the odor of others.'

You shall love the taste of others, flavour of the other's flesh: ethic as cannibalism; love itself parasitic on the violence of incorporation. You might love this. You will love this.

Inversion 4: Grant's plan reverts the Oulipian system. For while the replacement of every X noun keeps the signifier-signified relation stable while arbitrarily altering and substituting signifiers, Grant arbitrarily alters the signified of the infected signifier, appearing, at the moment lips promise themselves to each other, to land on Ludwig Wittgenstein's theory of the meaning of a word as neither more nor less than its usage.

Does the film thus suggest aesthetic compensations for violence, allowing *new words* (we are tired of *hosting* the old ones) to halt the devastations of mass infection? (Inversion? The poetic for the pestilent?) No. For Grant's attempts to redeploy words are ultimately futile. Despite one final broadcast pleading for people to *Stop* – understanding, killing, making sense – a military voice counting from ten obliterates the world of the diegesis in the cut to black at the number one. The problem all along was that disinfecting individual words did not obliterate the speaking of language. '"Kill" is kiss' does not kill the word; reciprocal determinations are not only still possible but are structurally necessary. Grant's undead problem, though perhaps unanticipated by Jacques Derrida (1997, 158), is accordingly given a perfect account in the most famous passage from *De la grammatologie*:

> Yet if reading must not be content with doubling the text, it cannot legitimately transgress the text toward something other than it, toward a referent (a reality that is metaphysical, historical, psychobiographical, etc.) or toward a signified outside the text whose content could take place, could have taken place outside of language, that is to say, . . . outside of writing in general . . . *Il n'y a pas de hors-texte.*

No novelty in the final term of *x is y* returns the real of the referent to *x*; there is no *y* after which there cannot be the deferral of another *z*; and *z* is not an end, but loops, and to an *a* that is never first or Adamic, prior to and securing 'is'. While horror's investment in the finitude of being seems generically self-evident, the more originary violence in this film is that of language, which is to say a violence that is fundamentally deployable *against itself*. For the only way to grapple with the finitude of comprehension is to develop new language, promise the infinity of potential discourse – but one is, as one is, always and still stuck with some message. ('As you know inoculation is the weapon of choice against virus and inoculation can only be effected through exposure' [Burroughs 1967, 10].)

Infection of the host is no longer a difference to the self, for one was already a carrier: compromised because the subject is the subject of language, infected in advance so that one might speak at all. Invaders (viral things) invade the invaded (mouthy things). Not devastation or radical compromise: the virus merely adds – and almost imperceptibly – to the world's din.

Rotation: Derrida's axiom, slightly permuted – *Il n'y a pas de hors-mot.*

Rotation, second effort: Il n'y a pas de hors-bruit.

Pontypool is thus an infection film with no possible disinfectant – as opposed to the self-reflexivity of much postmodern horror, this is a *self-referring* horror film. Replication's violence simply goes on; that is what (all) it does. The exteriority or space of transcendence that signals the apocalyptic breaking into normality in other zombie films (from the crackles of the radio in *Night of the Living Dead* to the flatline television signal in *Shaun of the Dead*) is here correlated fully to the mode of the virus – for even noises, and static, those gestures that in other texts signal failures or *withdrawals* of language, are ultimately co-opted here by language. The problem, of course, of any virus is one of repetition, replication, reproduction; as Steven Shaviro (1995, 40–1) writes, the virus is a message 'whose only content is an order to repeat itself . . . Language is one of these mechanisms of reproduction. Its purpose is not to indicate or communicate any particular content, but merely to perpetuate and replicate itself'. Is noise, here, then, or the radio static that dominates the soundtrack as beings fall away, the true infective force? Does that static infect and thus pervert, permute the viral replications of language, occluding the perpetual announcement of the something-that-is-happening?

In *The Parasite*, his fable-drenched demonstration of irreversible, unidirectional relationality, Michel Serres (2007, 6) notes that, in noise, '[o]ne parasite chases another out. One parasite (static), in the sense that information theory uses the word, chases another, in the anthropological sense'. Serres will write of his titular term, literally 'to eat next to' (*para*: near; *sitos*: grain, wheat, corn; – food, thus sometimes the flesh of the neighbour) that the parasite compromises the integrity of its host, inflames, makes such noise. Parasitic relationality names the logic of language, violence and love in *Pontypool*: consider it another fable for the philosopher; – *yet the undead are not the rats.*

The rats were always words.

Serres' parasite is both a relation and a form: figuring host to guest, house to intruder, system to disturbance, it also names minor difference, irrevocability, is the 'thermal exciter' of systems – it points, deictic bug, then, both to every modification to relation and to the relation itself. Thus, his use of *parasite* as, at once, concept, and the noise that undoes the concept (what *affects* the concept: in other words disturbs; in other other words, converts the abstraction of concept to the register of affect) shifts between the multiple registers that are also operative in *Pontypool*: as biological (the virus), as social (mass violence). But Serres also focuses on how the French for *parasite* connotes static, noise on the line that contaminates the message, noise as interference. The parasitism that feeds on the speaking subject is coextensive with the crackling radio sign of decomposition. Static renders the message incomprehensible, signals interruption and variation. But while all this 'excitement stops the message from passing', it also sometimes 'allows the message to pass, a message that cannot cross an unexcited channel . . . White noise is the condition for passing (for meaning, sound, and even noise), and the noise is its prohibitor or its interception. Noise, or again, the parasite, is at the three points of the triangle: sending, reception, transmission' (Serres 2007, 194). Getting rid of the parasite (devastating the virus) means getting rid of communication. In *Pontypool*, this noise, and also static, this spreading viral language, is coextensive with death, with the most terrible violence. Hence the film's tagline 'Shut Up or Die' *lies*; there is no shutting up or out of language sufficient to avoid the consequence that the 'or' promises. The film, in deploying language as the formal motor of violence, thereby structurally does not admit a space in which violence could be *said* to cease.

The infected are to the letter parasites: feeding off others, they give nothing in return; except: they give everything in return: they give the virus in return for the body as food. Feeding on (feeding next to) they also feed beside (as in: in addition), feeding film form on, consuming in order to generate the very ongoingness of the text. Formal language is thereby the most ravenous parasite of all.

One consequence: This is a horror film that eschews the rhetoric of misrepresentation or absurd amplification that is the privileged eidos for so many horror texts (in, say, the excessive de-monstrations of the monster). *Pontypool* does not exaggerate how signs work in language: rather, it displays, brutally presents, diagrams – as though giving nothing but a topology – what is constitutive of language (metonymic displacement slides; the autonomy of the signifier; mobile interconnectivity; dynamism, replication, transmission). This failing to fall to exaggeration is another mode of the film's stark simplicity, that it is, in some ways, cinematically coextensive with language. *Pontypool* is thus not a representation, but it is akin to an *attestation*. And what it attests to, what it bears witness to as its representational form, is the violence of referral that is, at once, how language works, how infection spreads, how bodies fall, one after the other. And – perhaps with the least exaggeration of all, which is to say conforming to what is most constitutive – all its vocables also attest to the noisiness by which love worms in.

Last point of emphasis: This day in question, violence's day? Valentine's Day.

Then what of the difference between noise and static?

Noyse, from the old French for din, disturbance, clamour, outcry; perhaps from *noxia*, hurting, damage, perhaps from *nausea*, disgust, discomfort (literally, seasickness: a sickness of and on movement); *chercher noise*, to pick a quarrel, an obsolete sense of rumour or report. *Statikos*, *statica*, from *histanai*, causing to stand, pertaining to the science or art of weighing, a placing in balance; having to do with bodies at rest or the production of an equilibrium; later, random or hissing radio noise, electrical interference; even later, aggravation, angry criticism, words out of balance, ones that are not terms of endearment.

I tell you this is what dying is –

Deferral is referral (or: the amative infective)

Right at the film's midpoint, *Pontypool* breaks. The violent throng has finally arrived at the threshold of the station; they pronounce their infectedness in an echolalia of words shouted by the protagonists into the cold air. Now, there is nothing to do but barricade and go back on the air. Obituaries are next, and with nothing taped, Grant goes live. The morning show is now a mourning show. The film cuts from its mobile ever-panning camera and colour stock to overexposed, strangely frozen black-and-white images (they are not photographic stills; vitality slightly pulses in the images), all details obliterated in flat stark planes of facticity as the voice declares that these beings are now dead. The image flickers, redirecting static from the sonic to the visual, and, as in Serres' account of parasitic-static, it interrupts and intrudes on the image while simultaneously sending its message that something has come undone here. Over changing images, stark and bleached, blinks and twitches, two girls on a street, a figure against a blank background, then another, now a man, another then and so on, the voice flatly intones, not always aligning with the images:

> Gwendolyn Parker was taken from this life in her 45th year by her beloved husband Stanley, who left this world suddenly, at the hands of family members Fiona and Michael, who then died at each other's hands in their 12th and 17th years respectively. Janice Gwynne has departed from her abiding husband, and by his own hand, in the 34th year of her life. Jack Gwynne survived long enough to add four names, Paul Heighton, 43, Alice Heighton, 42, Brenda Heighton, 12, and young Jesse Heighton, 10, to a list of passages before himself losing his life as a result of an accident. Greg Owen, 56, has been killed by Yolanda Owen, 61, who also removed Frieda Owen, 81, Patsy Owen, 12, John Freethy, 33, Peter Stamp, 38 and Leslie Reid, 42, who had, between them, caused the untimely passings of Joel Froth, 67, Sandra Weydon, 23, Tim Drummond, 17, Cynthia Drummond, 46, Darren Drummond, 51, and Alicia Drummond, 91. The Drummonds were survived on Cynthia's side by the Hindman family, until shortly before noon today when they were sadly removed from this world by a bus driven by the recently departed Brenda Lockland, 43, who was missed, briefly, by her husband, Gary, 37, now deceased.

The film's limited setting, small cast of the living, and aesthetic rules for mobility and colour here give way in a caesura that signals a leap – and the rhetorical mode that cuts across these stark images of the relentlessly deceased is a formula of *referral.*

Minor distinction: Deferral delays (scatters, disperses, postpones); referral redirects (relates, attributes, assigns).

Each survival in the monologue traces back to an obliteration, each devastation attributes a prior actor or future repetition; while the vocalization constitutes a proper obituary (a registry or record of departures) it also renders death a formal rhythm of relation and a brute assignation for disappearance. This chain of referrals ends arbitrarily and without *telos*; it is only *peras*, mere limit – which is to say its flat list just goes on no more. But before that, the referrals put on display the essence of what Maurice Blanchot (1988) and Jean-Luc Nancy (1991) figure as, respectively, the 'openness of a community' or 'the threshold of community': the witnessed deaths of others – what 'puts me beside myself' and 'calls me most radically into question', goes Blanchot's insistence in *La Communaute inavouable*. Because a final term for the listing is deferred through the form of referral, the catalogue of finitude colludes with fantasies of infinitude. Though the voice does eventually give out as it wears, the list form formalizes every future possible entry. The virus is thus not only a formal exciter narrativized in the stable box of the narrative: viral infection is the formal exciter of the text as such: modifies its univocality: creates change: generates the minor non-acclimatized aesthetic difference that is formal specificity.

Referral is given this formal *energia* through one overarching figuration: *beloved.* The list puts in relation the adored, the abiding, the familial, those who miss and are missed more when the other dies. What cuts across this registry of the dead – and the privileged form of viral vulnerability in language – is affection, endearment, care. *Love* is the third term, the disruptive interloper in the ur-form of violent relation: *killed by, killer of.*

When Nancy (1991) writes in '*L'amour en éclats*' of the problem of declaring love, it is the meekness of the repetitive declaration against its meant enunciation that makes a problem of speaking otherwise than exhaustively *about* love. Because philosophy has thought love as the 'extreme moment' that signals 'a being reaching completion', affiliating it with culmination, accomplishment, the final and definitive instant, but simultaneously as what exists beyond the self, the fundamentally dialectical dimension of love contains 'the contradiction of contradiction and of noncontradiction' (87). Philosophy in this way is always *missing* love: what is not there is not speakable by the philosopher (translated from affect into concept). Philosophy, itself nothing but love-of, always speaking *around* love itself. All noisy hiccups at the banquet.

Instead, Nancy writes that love arrives as its passage: 'love is the impossible, and it does not arrive, or arrives only at the limit, while crossing'. It is its coming-and-going, endless difference, excited flow. And Nancy (1991, 99) will call that which love unveils: finitude. 'Love does not transfigure finitude, and it does not carry out its transubstantiation in infinity . . . Love cuts across finitude, always from the other to the other, which never returns to the same.' What this means for words is that in saying the small but enormous, 'I love you', *nothing happens*. These words do not bring anything about; they name no thing, describe no substantive and deploy language in a manner that is unauthenticatable. Barthes (1978, 148–9) will likewise posit that the expression falls neither to linguistics nor to semiology, writing of the I-love-you that it 'has no usages' and is 'extra-lexicographical'. Serres (2015, 166) will likewise posit love as a recognition of what its utterance negates: '"I love you" doesn't specify place; "I love you" doesn't exclude anything, lets everything in, is ignorant of status, determination.' All stated I-loves/I-love-yous are promises, Nancy writes, but are always ones spoken in vain: 'We will have to maintain that love is always present and never recognized in anything that we name "love"' (93). *Something is always going to happen. Except in the I-love-you in which nothing at all is ever going to happen.* Excluding nothing, without use, it does not act, it does not. Serres' parasitic ur-form, in which 'the flow goes one way, never the other', the valve with no possible reverse direction, is the issuance of the amative without assurance. Without return; without response; without guarantee.

Translation, restatement: [I love you, beloved = {Ø}.]

Pontypool turns (and it is a non-arbitrary turn, for this is how its catastrophes go) on figures of this endearment because love, in being spoken *as its idioms*, and in thus making nothing happen, is a pure form of *language that does not fulfill itself*. Neither naming nor performative, the emptiness of these terms and these avowals are parasitic on all that has been named before while simultaneously effecting only the possibility that this nothing may also be a lie. If this film is about finitude and violence, parasitism and the brutality of infective stutters, it is also, and formally on the level of the lexeme, about endearment, about how much it is possible to say about just how impossible it is to say, about the limit of what it is possible to say, expressed in the utterance itself. 'The promise does not anticipate or assure the future', Nancy continues, 'it is possible that one day I will no longer love you, and this possibility cannot be taken away from love – it belongs to it.' Against this possibility – and let us say, perhaps even likelihood – love unveils the law of the word as both promise and limit. 'For one does not know what one says when one says, "I love you"', Nancy (1991, 101) concludes, 'and one does not say anything, but one knows that one says it and that it is law, absolutely.' What cuts across departures of being and language in the film is the leeching language of uttered endearment for the other. *Leeching*: as in bloodletting; as in oozing; what sheds and weeps, and also what profits.

Beyond fragile bodies (which the film refuses to subject to the finality of death through the meta-gesture of undeadening) and beyond viral language (which defers infinitely within its finitude), it is here that one finally arrives at the mode by which this horror film is grappling with what consumes all Horror-as-such on some level. But departing from the form's conventional obsessions with the *nothing* of non-being, *Pontypool* imagines a violence more terrible still, in a simplicity akin to another of Serres' (2007, 182) figurations of the parasite: 'One feeds on another and gives nothing in return.' So in declarations of adoring, so in care. To be clear: One says nothing with all the tender argot of affection. Nothing is made to happen when I say the words that I say when I say that I am saying that I love you. Terms of endearment are just so much static. And despite eschewing all graphic images of dismemberment, despite taking place entirely in the limited setting of a radio station (indeed, despite frequent critical accusations that *Pontypool* is the most boring horror film ever made) – in the most violent gesture possible, the film's formalization of horror is parasitic on expressions to the beloved.

This, to tell you, is what I'm dying for –

Coda: parasitologies

The uninvited guest.

Stupid, trivial, minor, popular.

Interrupts all labors of thought.

And weakens a discipline (without ever killing it).

What bristles the hairs (impudent!) on the neck of the philosopher.

Horror parasites philosophy: excites it; brings its shadowed bodies into a partial slanted shadowed light.

The arrow here is the bite. The flow that goes one way (and never the other) is the deadly infection. 'Man is a louse for other men.'

But this is not a matter of metaphors.

Horror feeds off philosophy *and gives nothing in return.*

Horror literalizes – which is to say: acts without exaggeration; demonstrates as the necessary – a relation of giving to the other: death. And new words. They function as the same thing. It is not a matter of instrumentalizing the philosophical monstrations (nor their monsters) as an allegory minable in or for the film: the film thinks this relation; it gives the relation form. Or the relation gives the horror its form.

Either way, the parasite is not a trope. (Operations are not figurative.) All horror is consumed with the parasitic question; *The Parasite* gives it many times, once as: 'What living together is. What is the collective?' (These are fundamentally questions of violence. Or love. Mostly violence.)

Serres specifies the problem with thinking these matters: '[T]hey do not say distinctly enough whether they are a philosophy of being or of relating. Being or relating, that is the whole question' (2007, 224). Restatement: ontology or ethics, that is the whole question. (For horror, that is the only question.) Where does violence reside, a violence integral to being or a violence integral to relating: is the violence of finitude *my* death above all else – nothing else individuates *me* – or is it in fact the death of the soft vulnerable other? Horror is sustained by that difference: *it feeds off that difference.* Horror is thus quasi-philosophical, in the sense of Serres' quasi-object: it designates a missing piece neither speakable within philosophy (affect missed by concept) nor entirely apart from it; horror names the *or*: what unsettles, refuses to settle: Being *or* relating? My death *or* yours; or an emergent third death to which only form can speak, death in referral, death in the chain that flows one way, irrevocable, irreversible, deaths that are hosted by the form of a sequence? Serres dubs the parasitic the 'single arrow, this relation without a reversal of direction' (5). Man hosts the other (grieves, retains, memorializes, protects); man hosts the other (houses the virus that replicates, is the source of infection): each survives only to be survived by another.

One only ever bleeds *out.*

References

Bacon, Simon. (2015). Bad Language. In: Nadine Farghaly (ed.), *Beyond the Night: Creatures of Life, Death and the In-Between.* Newcastle upon Tyne: Cambridge Scholars.

Barthes, Roland. (1977). The Death of the Author. In: Stephen Heath (ed. and trans.), *Image, Music, Text.* New York: Hill and Wang.

Barthes, Roland. (1978). *A Lover's Discourse*. Trans. Richard Howard. New York: Hill and Wang.

Bataille, Georges. (1985). Eye. In: Allan Stoekl (ed.), *Visions of Excess: Selected Writings, 1927–1939*. Trans. Allan Stoekl et al. Minneapolis: University of Minnesota Press.

Blanchot, Maurice. (1988). *The Unavowable Community*. Trans. Pierre Joris. Barrytown, NY: Station Hill Press.

Burgess, Tony. (1998). *Pontypool Changes Everything*. Toronto: ECW Press.

Burroughs, William. (1967). *The Ticket That Exploded*. New York: Grove Press.

Burroughs, William. (1998). Nova Express. In: James Grauerholz and Ira Silverberg (eds), *Word Virus: The William S. Burroughs Reader*. Intro. Ann Douglas. New York: Grove Press.

Deleuze, Gilles. (1997). He Stuttered. In: *Essays Critical and Clinical*. Trans. Daniel W. Smith and Michael A. Greco. Minneapolis: University of Minnesota Press.

Deleuze, Gilles and Félix Guattari. (1986). *Kafka: Toward a Minor Literature*. Trans. Dana Polan. Minneapolis: University of Minnesota Press.

Deleuze, Gilles and Félix Guattari. (1987). *A Thousand Plateaus: Capitalism and Schizophrenia*. Trans. Brian Massumi. Minneapolis: University of Minnesota Press.

Derrida, Jacques. (1997). *Of Grammatology*. Trans. Gayatri Spivak. Baltimore: Johns Hopkins University Press.

Heidegger, Martin. (1971). *Poetry, Language, Thought*. Trans. Albert Hofstadter. New York: HarperCollins.

Nancy, Jean-Luc. (1991). Shattered Love. In: Peter Connor (ed.), *The Inoperative Community*. Trans. Peter Connor, Lisa Garbus, et al. Minneapolis: University of Minnesota Press.

Serres, Michel. (1997). The Geometry of the Incommunicable: Madness. In: Arnold I. Davidson (ed.), *Foucault and His Interlocutors*. Chicago: University of Chicago Press, 36–56.

Serres, Michel. (2007). *The Parasite*. Trans. Lawrence R. Schehr. Minneapolis: University of Minnesota Press.

Serres, Michel. (2015). *Rome: The First Book of Foundations*. Trans. Randolph Burks. London: Bloomsbury.

Shaviro, Steven. (1995). Two Lessons from Burroughs. In: Judith Halberstam (ed.), *Posthuman Bodies*. Bloomington: Indiana University Press, 38–56.

6

The exogenesis of light

Matteo Pasquinelli

In other words, the atoms encode. Material elements, they perform as well as signs; they inform each other mutually, elect each other, choose each other, reflect each other, repel each other, like the diamonds in that cave of wonders, like all molecules do, like the codes of the living combine with each other and eliminate each other . . . They encode, we encode; they count, we count; we speak, they speak. Knowledge is thus the ability to listen and to translate the scattered languages of things. They usually speak mathematics.
– Michel Serres, 'Information and Thinking'

Computing in the dark

In his 2017 essay 'Information and Thinking' Michel Serres presented the scintillation of darkness as a better archetype of knowledge than the worn-out metaphor of *enlightenment*. Serres (2017, 13–20) counters what is for him the cruelty of daylight abstraction with a serene nocturnal cosmology, in which a multitude of stars displace the Sun King of Western reason from the throne it has occupied since the 1700s: 'More beautiful than the day, peaceful by all means, the star-studded, pensive and soft night is a better model of knowledge than the sun-struck, cruel, exclusive, eye-hurting, ideologically-prone and opinion-ridden light of day.' Plato's cave is the canonical allegory that has been repeatedly singing only 'the glory of one sun'. In order to illustrate his own counter-allegory Serres refers to a prismatic grotto found in Jules Verne's (1911) novel *The Star of the South*. Following the tradition of *plural materialism* that is typical to French philosophy at least since Gaston Bachelard, Serres descends into Verne's underworld to implode 'the light of reason' into a myriad of prisms:

> They were in the center of an immense grotto. The ground was covered with fine sand bespangled with gold. The vault was as high as that of a Gothic cathedral, and stretched away out of sight into the distant darkness. The walls were covered with stalactites of varied hue and wondrous richness . . . The decomposition of the luminous rays by the thousands of prisms, the showers of brilliancy that flashed and flowed from every side, produced the most astonishing combination of light and color that had ever dazzled the eyes of man. (276–7)

What is the relation between the web of flickering rays described by Serres and the current technosphere of digital networks? In the same lecture (see the quotation at the opening of this text), Serres attempts ambitiously to unify ancient atomism and modern computation in what looks like a gentle and moderate version of Gottfried Wilhelm Leibniz's *mathesis universalis*. Serres conflates the different dimensions of energy, light, information and knowledge into the figure of a cosmic computer with no central algorithm: neither a central sun nor a central human directs it.[1] Such a cosmology clearly responds to the historical juncture of computation: sensing the current technical composition (i.e. the acceleration of intelligent machines), Serres aims to bring atomism to a computational level (attributing computation also to nature) and, simultaneously, to disperse computation into the infinite universe (depriving therefore the Turing sphere of its epistemic and political hegemony).

Such computationalism is as a new version of atomism rather than of vitalism. In fact *pancomputationalism* (the idea that 'everything computes') must be distinguished from the doctrine of *panpsychism* (the idea that 'everything thinks'). From cybernetics to neuroscience, from philosophy to physics, the idea of computation appears to be used more and more often today to challenge and secularize the conceptual hegemony of both the living and the thinking. Computation emerges as the profane ground that aims to unify the two traditionally separated domains of *res extensa* and *res cogitans*, the sciences of nature and the sciences of spirit.

Curiously, despite his critique of the Enlightenment, for Serres the medium of universal computation is once again *light*, still wearing, like the god Hermes, the winged shoes of the 'third messenger' between matter and cognition.[2] To rethink a politics and aesthetics of light it would be interesting one day to reverse this correlation. Will darkness ever have its own medium of communication? Will it ever be possible to envision a medium that operates via negation, abduction, absence, the void and the non-luminous? Is the darkness able to compute?

The ultimate capital is a databank

The first to suggest that the cosmos may be a calculator was the Berlin-born engineer Konrad Zuse, a pioneer of modern computer and programming languages, in his 1969 book *Calculating Space*. For Zuse, the entire universe could be described as the output of a deterministic computation of a single *cellular automaton*. Half a century later, however, Serres' universe still resembles, curiously, just the very basic information channel articulated by Claude Shannon and Warren Weaver's model with no recognition of the progresses in cognitive sciences, artificial intelligence and machine learning: 'Bacteria, fungus, whale, sequoia, we do not know any life of which we cannot say that it emits information, receives it, stores it, and processes it. Four universal rules, so unanimous' (Serres 2017, 13). Also trees compute, adds Serres, as they 'calculate' each year in the rings of their trunks. We may like the idea that the whole of the universe computes and scintillates, that forests calculate, that mushrooms keep on extending the Earth's natural internet, yet how can these cosmologies provide a basic grammar of ecology and economy that is useful to explain the crucial phenomena of surplus, capitalization, crisis and transformation?

In *The Parasite*, first published in French in 1980, Serres (1982a) took the sun as the symbolic form of capital par excellence: 'The ultimate capital is the sun', he wrote, retracing the history of ancient solar cults and their philosophical reincarnations. The *political economy of the solar system*, namely, the coupling of terrestrial photosynthesis with sunlight, was probably introduced into French philosophy by Georges Bataille.[3] Unlike Bataille and his idea of sacrificial expenditure, Serres related the metabolic accumulation of solar energy also to the tendency of terrestrial capitalism towards abstraction. Serres' idea of energy included the dimension of language and the progressively abstract flows of money, signs and data: not just the transformation of energy into money and money into signs, but also the transformation of information into data and the rise of 'databanks'.[4] Serres was writing in 1980 and such a form of capitalization – the accumulation of information on information – can be properly understood only today, in the age of the metadata society and the global datacentres of Google, Facebook and the NSA (see Pasquinelli 2015, 49–68).

> What is capital? It is the reservoir above the dam, an iron mine or a coal, manganese, or tungsten mine; a gold mine. An oil well. It is a stock of energy and of primary material; it is an island of negative entropy. It is a store of writings. The old standard of precious metal, having become banal, tends to disappear. We are moving toward a data bank. These reservoirs are only subsuns. Their source, far upstream, is the sun. The real, ultimate capital is the sun. (Serres 1982b, 173)

In the same passage Serres criticizes Bataille's philosophy of excess as just another 'subcult of the sun' that still falls within the conventional worship of our regional star. Indeed Bataille belongs to the canon of modern philosophy and its intrinsic *heliocentrism*, or the idea that the sun (clearly a *doppelgänger* of capital, according to Serres) would be the centre of all worldly affairs.[5] Whereas Bataille was providing a model of surplus value (albeit a vitalistic one), Serres appears to be in fact a materialist philosopher without a model of surplus value and accumulation – that is, a basic grammar of transformation. Serres' figure of the *parasite* was an attempt to replace Karl Marx's notion of surplus value with a different ethical profile to cover both the domains of ecology and economy, but still it remained on the same level and scale of the agents that it was supposed to parasite. Curiously, the diagram of the parasite sketched by Serres in 1980 was still drawn according to Shannon and Weaver's communication model (sender–message–receiver). In French, the word *parasite* also means the noise of static electricity in an information channel such a telephone wire: Serres' idea of the parasite is without doubt still the grandchild of the second law of thermodynamics, or the law of entropy.

Liberating information from entropy

At a certain moment in the history of thought, information came to be equated with energy as a form *negative entropy* – by imitating precisely the definition that was used for the photosynthentic process in charge of transforming sunlight into the high-energy molecular bonds of sugar. This happened in a curious way, by equating the negative descriptions of the two measures: as the formula of thermodynamic entropy (the gradient of heat irradiation) showed similarities with the formula of informational noise (the level of randomness in a communication channel), it was deduced that energy and information were more or less the same thing, or that they were sharing the same ontological *continuum*.

Informational entropy speaks of information as a quantity, not as quality. It says nothing of semantic information. The two mathematical formulas may be similar, nevertheless semantic information belongs to a different domain than energy, and so computation. This confusion affected philosophy, media theory and the arts – an example of the generalization of scientific theories and application to the most disparate domains (notoriously including also psychology and economy). How was this misunderstanding born?

Entropy is a measure of disorder within a system: that is, of how much energy is 'disorganized' or in chaotic form. This notion was introduced by the German physicist Rudolf Clausius (1867), who encapsulated it in the formulation of the Second Law of Thermodynamics: given the energy of the universe remains constant, the entropy of the universe tends to a maximum. For example, the entropy of a room is said to increase if furniture is set alight and the energy enclosed in the matter is transformed into heat and dispersed by combustion. The Second Law of Thermodynamics states that the energy disorder of any closed system tends to increase and leads to a uniform equilibrium: it says that everything decays and especially that heat tends to irradiate and dissipate. From the sun to violent geological events, the *inorganic world* shows clear phenomena of entropy: it burns energy and increases the chaos of the universe.

In the 1940s Claude Shannon and Warren Weaver defined 'information' as the opposite of noise. In quantitative terms Shannon decided to measure the quality of communication as a negative value against the background noise of a given channel. The lower the noise on the channel or 'information entropy', the higher the quality of communication. The application of the term 'entropy' to 'information' was suggested apparently by John von Neumann. Shannon did not know what to call his mathematical measure of information:

> My greatest concern was what to call it. I thought of calling it 'information', but the word was overly used, so I decided to call it 'uncertainty'. When I discussed it with John von Neumann, he had a better idea. Von Neumann told me, 'You should call it entropy, for two reasons. In the first place your uncertainty function has been used in statistical mechanics under that

> name, so it already has a name. In the second place, and more important, nobody knows what entropy really is, so in a debate you will always have the advantage.' (Qtd in Tribus and McIrvine 1971)

The term 'informational entropy' provoked some confusion as it suggested that information processes negate the Second Law of Thermodynamics: on the contrary, *information order* in fact does not accumulate energy out of chaos. A virtual database, to give a surreal example, cannot 'burn' (even though energy was needed to compute it). A memory stick full of structured data contains the same amount of energy as an empty one made of a meaningless sequence of 0 bits (even though more energy was needed to compute the former). Digital impulses can be transmitted perfectly today via very noisy channels. Shannon's homology of information noise and entropy was audacious and its consequences are still visible today. The misunderstanding has also other roots.

In his 1944 book *What Is Life?* Erwin Schrödinger, observing cellular metabolism and the exchanges of energy between the inside and the outside of the cellular membrane, concluded rightly that life bypasses the second principle of thermodynamics – the law of entropy. Quite the reverse, aside from consuming energy, cell metabolism is also able to accumulate it. He wrote:

> Every process, event, happening call it what you will; in a word, everything that is going on in Nature means an increase of the entropy of the part of the world where it is going on. Thus a living organism continually increases its entropy, or, as you may say, produces positive entropy and thus tends to approach the dangerous state of maximum entropy, which is death. It can only keep aloof from it, i.e. alive, by continually drawing from its environment negative entropy which is something very positive as we shall immediately see. What an organism feeds upon is negative entropy. Or, to put it less paradoxically, the essential thing in metabolism is that the organism succeeds in freeing itself from all the entropy it cannot help producing while alive (Schrödinger 1944, 71–2).

Photosynthesis transforms solar energy and stores it in carbon rings of sugar and cellulose. This flow of energy feeds the whole ecosystem up up to predatory animals and even the civilization of machines ('fossil fuel' like oil was living matter once). Going *upstream*, this flow of energy continuously challenges the law of entropy, which is the tendency of the mineral world to dissipate energy. Schrödinger crystalizes the enigma of life itself in the idea of *negative entropy*. Importantly, he recognizes life as a process of *positive accumulation* and not just as a *negative resistance* to entropy.[6]

Eventually the term *negentropy* was coined and applied by the French physicist Léon Brillouin to information theory, probably in this way originating a *liaison fatale* between energy and information that has been absorbed by the whole of French philosophy.[7] For Brillouin the processing of information always requires energy, and he attempted to measure how much energy is required to produce a simple bit of information. Producing negative entropy of information in one region of the universe means raising the entropy in another. Yet informational negative entropy and thermodynamic negative entropy refer to two completely different scales and should not be confused or considered homologous. Information, and specifically semantic information, has nothing to do with energy, even though energy is required to compute information. This conundrum resurfaces in the work of Gilbert Simondon (1958, 16) as well:

> At the level of the technical ensembles of the twentieth century, thermodynamic energeticism is replaced by information theory, the normative content of which is eminently regulatory and stabilizing: the development of technics seemed to be a guarantee of stability. The machine, as an element in the technical ensemble, becomes the effective unit which augments the quantity of information, increases negentropy, and opposes the degradation of energy. The machine is a result of organization and information; it resembles life and cooperates with life in its opposition to disorder and to the levelling out of all things that tend to deprive the world

> of its powers of change. The machine is something which fights against the death of the universe; it slows down, as life does, the degradation of energy, and becomes a stabilizer of the world.

This is of course Simondon's famous critique of the Mathematical Theory of Communication (MTC). Simondon however could not forecast the vertigo of contemporary computation, the alliance of the human with a new generation of intelligent machines, the frequent bugs and crashes of computational capitalism (see the 2010 Flash Crash, for instance). Intelligent machines and algorithmic governance are not always actors of stabilization, quite the opposite.

It may be wiser to frame the riddle of information and energy from a historical point of view too, and to look at the influence of the technological composition on cognitive scaffolding. The first massive *violation* in the terrestrial energy cycle occurred, for sure, in the eighteenth century with the introduction of the heat engine that launched the industrial revolution. A heat engine is a device that converts thermal energy to a mechanical output, nevertheless burning and dissipating more energy than what is actually transformed. Industrial machinery is designed to execute work and release energy in a constant and controlled flow – indeed, technology is a sort of *domesticated entropy*. In terms of their energy balance, heat engines are closer to the inorganic world than to living matter: they are simply able to consume those carbon compounds, such as coal, in which natural processes has slowly condensed energy. Thermal engines devour far more energy compared to natural metabolism and devour nature itself: a few centuries on and industrial by-products have visibly altered the biosphere to the point where they represent clear markers of the so-called Anthropocene age.

In the twentieth century, the information revolution introduced a further entropic regime. Electronic media and digital computers consume very little energy in comparison to industrial engines and have proportionally a lower energetic impact. More precisely, a Turing machine, being an abstract machine counting binary digits, is not related to a material substratum and theoretically consumes no energy at all: it runs ideally in a virtual space at almost zero entropy. From the perspective of abstraction, digital networks are purely mathematical spaces, with no gravity, no friction, no entropy whatsoever. The ethics and aesthetics of the digital, the abyss of the infinite reproduction of copies, the massive devalorization processes of the digital economy are all results of such an almost zero-entropy regime. Postmodern and speculative philosophies, too, can be seen as cultural reactions to the abyssal nature of a digital *logos* accelerating and running away at almost zero-entropy.

The exogenesis of light

How does energy eventually turn into information, and information into capital? How do these domains speak to and translate into each other? And, more importantly, how can we register forms and degrees of accumulation and surplus across the different levels of computation? Any neomaterialist and neorationalist philosophy has to confront this issue. The scale of current planetary computation inaugurates a new epistemic space that permanently changes Serres' old atomist points of reference and informational metaphors. From climate change to financial trading, from dataveillance to logistics, everything is computed across an infinite datascape encaged in the data centres of media monopolies. Mass computation is a form of power and capitalization that today sits alongside the traditional forms of economic, military and political power and it has come to constitute a new hybrid imperial *nomos*.[8]

It is only via a planetary scale of computation that climate change, the Anthropocene paradigm and even the Gaia hypothesis can be described, as Paul Edwards has shown in his fundamental book *A Vast Machine* (2010). The planetary network of climate science institutions incarnates a centralized form of computation that questions the political autonomy of the ecological paradigm. The 'solar databases' described by Serres in 1980 are here reversed: they do not measure and condense solar value, but they compute and politically contain the energy surplus of the Earth.[9]

The idea that living metabolism is based on the accumulation of energy against entropy (or *negentropy*) has been recently contested by new theories. Jeremy England at MIT has proposed new mathematical formulae to suggest that the evolution of complex organisms and species may actually be functional to dissipate *more energy* compared to the gradient of irradiation of the inorganic world.[10] According to this theory, life may have emerged on Earth beneath the continuous and excessive irradiation of sunlight: the 'pressure' of the Sun would have pushed molecules to form more complex structures in order to channel and disperse energy more efficiently. The multiplication of different species, and evolution itself, would be a more efficient strategy to broadcast energy and not just to accumulate it.

Although still in a controversial formulation, the idea that living matter transmits energy faster and more successfully than the old bricks of the cosmos may be useful to crack up the old vitalist diagrams in which nature is usually domesticated. Gilles Deleuze and Félix Guattari tried themselves to frame life as a *line of flight* pulled by the outside rather than as a drive pushed by an internal force. In their ontology, in fact, *exogenesis* replaces the old model of *endogenesis*. This is why the concept of *deterritorialization* logically comes before that of *territorialization* in their history of capitalist evolution. It is the outside which generates and drives the system and not simply the organism that projects and inhabits its own *Umwelt* (like in the German *Naturphilosophie*). What is fascinating in Jeremy England's hypothesis is that even the human mind and its form of extended cognition could be seen as an extension of this quest towards more complex architectures of energy. In this view nature is not simply an organization that emits, receives, stores and processes information: energy and light shape this very living matter from the outside and they innervate it to multiply exponentially. It is mere fatalism to leave this idea of *overgrowth* and irradiating complexity solely to the description of solar capitalism. Politics and aesthetics have to be part of this quest towards more complex architectures of light.

References

Bailly, Francis and Giuseppe Longo. (2009). Biological Organization and Anti-Entropy. *Biological Systems* 17.1.

Brillouin, Léon. (1949). Life, Thermodynamics and Cybernetics. *American Scientist* n. 37.

Brillouin, Léon. (1950). Thermodynamics and Information Theory. *American Scientist* n. 38.

Brillouin, Léon. (1956). *Science and Information Theory*. New York: Academic Press.

Clausius, Rudolf. (1867). *The Mechanical Theory of Heat: With Its Applications to the Steam Engine and to Physical Properties of Bodies*. London: John van Voorst.

Edwards, Paul. (2010). *A Vast Machine: Computer Models, Climate Data and the Politics of Global Warming*. Cambridge: MIT Press.

Negarestani, Reza. (2010). Solar Inferno and the Earthbound Abyss. In: Pamela Rosenkranz (ed.), *Our Sun*. Rome: Istituto Svizzero.

Pasquinelli, Matteo. (2015). Italian Operaismo and the Information Machine. *Theory, Culture & Society* 32.3: 49–68.

Prigogine, Ilya. (1977). *Self-Organization in Non-equilibrium Systems*. New York: Wiley.

Schrödinger, Erwin. (1944). *What Is Life?* Cambridge: Cambridge University Press.

Serres, Michel. (1982a). *The Parasite*. Trans. Lawrence R. Schehr. Baltimore: Johns Hopkins University Press.

Serres, Michel. (1982b). *Hermes: Literature, Science and Philosophy*. Baltimore: Johns Hopkins University Press.

Serres, Michel. (2000). *The Birth of Physics*. Trans. Jack Hawkes. Manchester: Clinamen Press.

Serres, Michel. (2017). Information and Thinking. In: Rosi Braidotti and Rick Dolphijn *Philosophy after Nature*. London: Rowman and Littlefield International, 13–20.

Simondon, Gilbert. (1958). *Du mode d'existence des objets techniques*. Paris: Méot. Translation by Ninian Mellamphy: *On the Mode of Existence of Technical Objects*. University of Western Ontario, 1980.

Tribus, Myron and Edward C. McIrvine. (1971). Energy and Information. *Scientific American* n. 224.

Verne, Jules. (1911). The Star of the South. In: Charles Horne (ed. and trans.), *Works of Jules Verne*, vol. 13. New York: Tyler Daniels, 276–7.

Wolchover, Natalie. (2014). A New Physics Theory of Life. *Quanta Magazine*, January 22.

Zuse, Konrad. (1969). *Rechnender Raum*. Braunschweig: Friedrich Vieweg & Sohn, 1969. English translation: *Calculating Space*. Cambridge: MIT Press, 1970.

7

Scintillant@the University of Angelic Invention

Gray Kochhar-Lindgren

[D]reaming of universities whose space is mixed and multi-colored, striped like a tiger, blended in different shades, dyed with numerous pigments, twinkling like stars – real like a landscape.

– MICHEL SERRES, *THUMBELINA*

An angel – although there is never just one – is a winged multiple who knows how to turn, how to swoop through vortices, how to look around with a thousand curious eyes and how to brush against a canvas, a code and a life. An angel knows the mobility of immobility and the stillness of the mobile. Wherever they appear and disappear, invention occurs. The laws of non-contradiction, the negative and critique shimmer on the tip of one of their wing-feathers as it glints in the sun. Angels, always close to music, muse and are musings. They enunciate annunciations; they bend close to the ear of all others. Composed, they compose. Messengers of the world, angels gather and disseminate; they transmit and transit. Angels do not exist, but perhaps, nevertheless, Michel Serres is an angel and perhaps the university is the head of a pin, a platform for a circuit board for transformative zettabytes.

Angels proliferate as image, sculpture, painting, language and code. Irreality, the virtual. A cloud of clouds. A word, passing in the night. The conceptuality of concepts and the possibility of the idea of the good. As an infinitely patient listening, watching and waiting in the library of the cosmos. Art, thought, invention: a gust of wind. What is happening with the event called the *university*? With its futurity? Venus – let's say the one painted by Sandro Botticelli – is approaching, ravishing but calm. Awaiting, looking ahead, the goddess of love is being blown ashore by a winged figure carrying a beautiful woman through the air. Their movement is blowing their own hair back behind them while their breath is blowing Venus's hair in the opposition direction, except for the long tresses that flow down her back and around her front so that her left hand, holding her hair, modestly covers the triangle of birth as she moves towards us out of the frame. Her right hand covers her heart.

The swell of the sea, blown by Zephyrus and Aura, whose knotted cloaks are exquisite, carries the shell on which Venus stands towards us from the ancient classical pasts and beyond us towards a futurity beyond what we have called 'modernity'. Venus is greeted on the shore by a Hora, a figure of the seasonal gates of heaven, who waits for the goddess, leaning towards her to drape her with a flowered gown. The university is the before, the beside and the to-come of the Botticelli, a new form of construction layered like an *impasto* – thick, congealed, reflective – and compressed into an image that must be read, always an act of unfurling in time, in order to be understood through the historical sedimentations, the delineations of a topology of an inventiveness that touches us profoundly.

Just behind and off to one side of Botticelli's *Birth of Venus* (c. 1485)[1] is the catastrophe of castration and warfare between father and son, one generation and its predecessor. *Polemos* provides the conditions for the history of the birth of love on an island in the sea, the air freshened by a fall of roses. The currents of the paint and the sea are surging, the wavelets breaking, the vectors moving in every direction. Serres will speak of the quasi-, parasites and the relationalities of *milieux*, prepositions and angels. This singularity of the always determined pluralized is uncontainable. The angelic is transmission, messaging, algorithms, the vectorial and the tranversal. Things break forth and the university is on the move.

Surely the angels – which do not exist – cannot save the university, for as we so often hear the university is finished: dissolving, decaying and crumbling into ruin. In our era of the wreckage of global financialization, the precaricity of labour, the brutality of terror, forced migrations, massive urbanization and the devastation of the earth as a dwelling place, the university – which has very nearly become completely delinked from its ideal of a development towards the whole (whatever this term might mean) – has become a cog in the R&D arm of corporate and state capitalism, a mirage of a past tradition of the arts in conversation with the sciences and a platform from which technology can ever more completely colonize both the macro- and the nano-logical interstices of the world. The increasing scope of the culture of the audit, driven by the wish for metrics to measure everything, is an index of this shift. Quantify and control for the sake of efficient production: thinking, learning and teaching judged by a delusional Taylorism run amuck.

The angelic as a (non)pattern of noise-information-noise, as a host of invisible messengers and as a gust from elsewhere cannot appear in metrics, but the question of the (im)measurable plays throughout the history of rationality, the exponentially expanding encyclopedia of knowledge, and of the university. Gaspare Polizzi (2000, 265) contextualizes this history in light of the ways in which Serres practices the transversality of writing, thinking:

> Reason, logos, and thought have their common genesis in the algebraic, static language of Greek proportion, in the stable equilibrium of the episteme (*TS*, xiv). Their most accomplished and concrete expression is the measure; sustained by the idea of always stabilizing an equivalence, measure is the symbol of the precision of scientific knowledge and of its legitimacy

> before reality, but it is also the metaphor of an unrealizable attempt to resolve reality into an equation, of the failure of incommensurability and indeterminism, of the excess of the nonmeasurable and of the immeasurable that is expressed in the vast contingency of individual experiences.

The excess of the incommensurable and of contingency are at work all the time, not as the irrational but as a different distribution of the rational plus a little extra, a reserve that provides the space of an auto- and heteropoetic inventiveness that opens along the edges of the university, which are everywhere.

The angels of Serres, like the work that prepositions accomplish, are relationalities that fly within middlings without fixed boundaries that cannot lend themselves to the appearances of phenomenality. Not as-such, although certainly as-if. The between of the milieu is imperceptible to the instrumentation of empiricism and the accountabilities of cash. The university, however, overrun by big data as a criterion of truth and by the corporation as a legislator of research, is, they say, breathing its last. Universities should be serious sites of useful production, not of the outworn modes of philosophical scholasticism, the uselessness of the arts and humanities and of the worn-out continuities of traditions.

All of these trajectories have been disrupted by the new titans of tech and the university should now be not just a factory for the industrial production of utilitarian knowledge, but also a start-up for the innovators, who are always in need of angel investors. Money from on-high, an invisible hand reaching down from the heavens to announce a new grant, a new demand. Angels, though, are themselves communicative technologies that always have to do with angles of flight, the play of light and number. How many dance? (All of them.) How does an infinity localize itself? (Constantly.) These messages and messengers, however, do not result from number and cannot be measured by metrics. This growing enforcement of metrics in our audit culture threatens angels not with non-existence – that has already been decided – but with irrelevance, which is even worse. If the angels indeed are vanquished, the university will be dead, without a reserve fund for re-inventing itself.

The Data Fetish that threatens thought, the inventive, and the idea of the university is a misapplication of metrics to the wrong category of object of experience. The metrical, taken as an indicator within the logic of quantitative, works well. Either an object is 2.5 inches long or it is not. Taken as a measurement of quality within the logic of the spectrum of the good and the not-good, as well as of the beautiful and its multiple others, metrics do not work. They are misplaced and misapplied. This is a very old philosophical conundrum laid out already by Plato, but it has received a fresh impetus with the multiple epistemological and ideological crossovers between global neoliberalism, the precision of technologies and the confusion between number, data and truth. But it is in this very space that Serres' work is inserted, a parasite on the body of learning, a virus coursing through the circuits of the motherboards and the boardrooms.

For metrics as the central category for the measurement of quality in the university, Jean-François Lyotard (1984, 4) had already pointed towards his future and our present with the prescience of *The Postmodern Condition*:

> The nature of knowledge cannot survive unchanged within this context of general transformation and it can fit into the new channels, and become operational, only if learning is translated into quantities of information. We can predicate that anything in the constituted body of knowledge that is not translatable in this way will be abandoned and the direction of new research will be dictated by the possibility of its eventual results being translatable into computer language. Along with the hegemony of the computer comes a certain logic, and therefore a certain set off prescriptions determining which statements are accepted as 'knowledge' statements.

More recently, Graham Allen (2016, 97), discussing the concept of 'transparency' in audit culture in the discourse of the contemporary university, claims that '[t]ransparency, as a dominating cultural value, equates accountability (to the government to funding agencies, to the tax-payer and the voter) with counting and with seeing . . . what exists is what can be counted (brought to account, accounted for) and what can be seen (assessed by audit, quality control, etc)'. The relationalities of Serres' angels, betweens and prepositions cannot be thus accounted for. One must, instead, historically juxtapose across vast stretches of time and space; one must see analogies across disciplines and methods; one must create a new map of knowing how to know. If the university is to breathe, we must read differently, compose differently, and redistribute practices, idealities, pedagogies and sensibility. Angels re-vivify.

One of the tasks of thinking in the dominant moment of audit culture[2] is to demonstrate the limits of the ideal of transparency. Simply put there is – much to the chagrin of the data fetishists and the hyper-technophiles – a necessary inescapability of the incalculable, the inscrutable, the reserve and that which does not come forth into perception, reason, good sense, quantity or knowledge. Serres, who has tremendous respect for the calculable and the measurable of metrics, calls this immeasurability by various names: Hermes, the between, prepositions and the angelic. Perhaps, then, we should not yet, not quite yet, write an obituary for the university, for the angels are always ecstatically immersed in the act of singing.

While it is the case that the university is battered about by the confluence of enormously destructive historical waves, this situation also gives us a hint of a possible opportunity, the slightest of openings, to learn how better to practice what Serres (2015, 19) has called the 'incandescent joy of invention'. What will we make of this chance, this hand of cards that contains an unknown number of jokers? What might the university become if it moves, in a slow and erratic process of burning,[3] towards the joy of a University of Angelic Invention? What, though, could be more absurd than such a ridiculous and nonsensical question?

Etymologies never provide grounds for rational proofs of an argument, but they do serve as openings for thought and as time-machines that whirl language from its distant past on the edge of hearability through the infinite reconfiguration called the present and towards a host of unforeseeable futures. The 'university' is an intriguing word. The '*uni-*' as a simple idea for the 'one' or the 'whole' has been spoken about enough for the moment, although we will always return to it. We are now more interested in the '*poly-*' or the '*multi-*'; assemblages, situations, archipelagos; edges, fissures, openings and porosities. Traces, spectres and filiations. The '*uni-*', we are beginning to understand, is an effect of the swarming micrologies of difference, which is not to dismiss the one – what would this mean for the angels? – but merely to reimagine its functions. In our dirges, eulogies and pleas for the university, we have not yet paid sufficient attention to the 'versus'.

The 'versus' of the university is a pivot, a turning of directionalities and orientations, a social *poēsis* that runs through *technē* and *physis*. The *versus* is a preposition and prepositions are, for Serres, sites of mediations, relations and relays. The word in its more distant past comes from 'PIE **wert-* "to turn, wind", from root **wer-* "to turn, bend" (source also of Old English *-weard* "toward", originally "turned toward", *weorthan* "to befall", *wyrd* "fate, destiny", literally "what befalls one"); Sanskrit *vartate* "turns round, rolls"' (Online Etymological Dictionary). The university is turning, but not on a stable perpendicular spindle, not on the axes of Euclidean or Cartesian geometry. It is off-balance, off-kilter and this inclination off of the vertical-horizontal axes grants it new options, a chance at difference. Lucretius is the Harlequin.

The university of invention always wobbles. It is a telecommunications nexus where a multitude of messages are exchanged, where an incessant texting is occurring that each of us (in a multitude of eaches) will interrupt with the inflections of our curiosity and the finitude of our limitations (both of which will be under constant revision by our immersion in a sublime typhoon of excessive signals). This university is a place where those who are 'able to give up the comforts of disciplinary specialism and risk putting themselves into perpetual translation' (Brown 2002, 12) will be able to construct provisional platforms of interactions with both the precise profundity of experts and the amorous eagerness of apprentices to Hermes with his invisibility, his quickness, his open secrecy and his acquaintance with both the skyways of the celestial gods and the sluggish dark waters of the underworld. The university will always be travelling, passing back-and-forth across its own boundaries in an action akin to the folding of proteins, unrolling the scroll of a painted map, the kneading of dough, the unfolding of a kimono or the swiping of a finger across the screen of knowledge.

Perpetual translation: always in motion, the impossibility of settling into place, swerving slightly or careening in a different direction, surfing. The emerging university of the *chaosmos* will, indeed, continue to demand extraordinary expertise across all the variations of the practices of discovery, but, even more essentially, it will require a slightly different disposition, a more torrential curiosity and, especially, a facilitation for making relays, detours, relations and mediations. The university to come – and it is already, in pockets, thriving here and there – will be the virtual-physical space of trying-out, of giving things a go. It will never be a stablizable utopian space, but, rather, a place of experimentation, of seeing what might happen *if*.

But, surely, angels have no place in the contemporary institution of rankings, research assessment exercises, grant funding, impact factors and teaching evaluations measured in decimals? If everything rests on the calculability of quality through a fantasy of the exactitude of numbers, then angels are banished forever. And, yet, angels move through the work of Serres like invisible but powerful bursts of signals move through interstellar space, through walls, through the earth. Since, however, we as moderns no longer 'believe' in angels, these winged messengers have withdrawn into film, advertising, video games and the kitsch of electronic greeting cards. We have forgotten what used to be known, as Polizzi (2000, n.2) reminds us, that at least from a Christian perspective, angelology is the

> theological science of angelic creatures. It draws its origin in Christian religion from the definition of the celestial hierarchy furnished by the Pseudo-Dionysus, with reference to Saint Paul . . . The angels are also present in the religious beliefs of the Old Testament and of Islam and assume the forms of breaths and vital spirits, or more often of superhuman beings who form God's court and are his ministers to men, to whom they announce His wishes. The figure of the guardian angel, diffused in Christianity, is also found in the Babylonian religion and in Zoroastrianism.

Angels, as paradox, are both visible and invisible. They appear in painting, scripture, literature and statuary; they appear in visions; the angel of history appeared, among others, to Walter Benjamin. There were, at one time, even rumoured to be angels in America.

And, yet, they all vanish in the twinkling of an eye, an *Augenblick*. This movement between the visible and the invisible, as nonsensical from a logical point of view as it is, is nonetheless a necessary aspect of their structural paradox. As Serres (HK Interview, 1995) reminds us:

> [T]he reason why angels are invisible is because they are disappearing to let the message go through them . . . If you read medieval angelology you find exactly the same demonstrations because all the problems for angelology – what is a message? who are the messengers? what is the messenger's body? – like Saint Thomas Aquinas, the early church fathers, the Pseudo-Dionysius, and so on. In the beginning of my book I quote the problem of the sex of the angels. Everybody smiles about this problem, but it is a serious one, a problem about transmission.

Angels are transmitters: they are messages and messengers; they are the noise that enables the logic of information and the networks of information sciences; they are the precision of *poēsis* that, mixing all things from the origin of origins, blurs boundaries between disciplines, between epistemologies, between students and faculty and between the university and its others.

Angels are everywhere. As the beings of the milieu of the between the wingbeats of their silent voices and their gravitational waves bind the world together into a disjunctive synthesis, a knowability of cause-and-effect, a song. The contemporary university of hypermodernity reduces, dissects and specializes. It recombines its particles, molecularities, fragments and departments. As a force of social power, it legitimates knowledges, positions, people and histories (which also always de-legitimates other trajectories). These are all powerful and necessary movements of global and local assemblages, but universities do not know enough about singing lessons of the soul, the wings of hummingbirds or about how numbers keep an unexpected beat.

Angels do not exist. Nowhere are they accessible to empirical verification or falsifiability. They are like grins without a cat; the strange ideality of ethics, mathematics or meaning; the individuality of a self that auto-affectively stretches between birth and death; the evanescence of the event of sense. They are also, however, prepositions, a dynamic connectivity of a work of nets that creates a university and its possibilities. And prepositions, in turn, are the 'algebra of fluxes' (HK Interview, 1995), angelic swarms of connectivity. Angels, for Serres, are the

> personification of prepositions. Here we use the angelic properties of prepositions as a means of making connections between two, between people and places, between theory and practice. Some prepositions emphasize position, the relation of an object or a subject to place, such as *on, in, between, through*. Others focus on relationships between subjects and objects, for example, among and with, and the directional nature of these connections like *beyond, for* or *to*. Yet others contain elements of time, as in *beyond*. (Rendell and Wells 2001; emphases in the original)

Beyond must be the most beloved of angelic prepositions, the one in and at which they laugh the most uncontrollably. What is it that exceeds every moment, every determination, and that cracks open towards the new? What's next? Without these prepositions there could not be a university (or an experience) of any sort, much less a university of angelic in(ter)vention.

Angels, though, are not only prepositions. They are the vast networks of digitized code; they are algorhythms-in-action. When, therefore, Serres speaks of an 'algorithmic' philosophy or we speak of such a university, we must keep in mind the multiplicity of topological vortices – a kimono, kneaded bread, crumpled paper – as well as the emergence of differentiated and unpredictable discoveries. The algorithmic in Serres is *not* the linearity of predictable recipes for behaviour or discovery. The procedure that moves in such a step-by-step manner is not thought, but only instructions for the redundancy of the logical conditions inherent in the set-up of the program. An algorithm is always, for Serres, an *algorhythm*, keeping a beat as its procedures create the folded rhythms of invention and discovery appropriate to new life on earth. The pulse, as pulse, exceeds the form of instructions to create the 'not-yet' and the 'next-space'.

In *Thumbelina: The Culture and Technology of Millennials* (2015), Serres rearticulates the culture of the university in the wake of the digital revolution by seeing different opportunities emerge for the power of number, coding and algorithms than the misapplication of metrics as a rule of quality. What happens, he wonders, when the space of the screen displaces the space of the book, around which the earlier forms of the university were organized? The old university, which still lumbers along most of the time in most places, was a 'pedagogical hierarchy reproduced in the focus of the courtroom on the judge, the theatre on the stage, the royal court on the throne, the church on the altar, the habitation of the hearth . . . of the multiplicity on the one' (33). In the traditional university the

> space of the lecture hall was designed as a field of forces whose orchestral center of gravity was the stage, with its focal point at the lectern, which

> was literally a *power point*. What was situated there was the heavy density of knowledge, which scarcely existed on the periphery. Now, knowledge is distributed everywhere, moving freely in a homogenous and decentered space. (34)

This space is unprecedented in its means of production and its opportunities – both destructive and creative – even if it has been prepared for aeons by Hermes and the angelic envoys; the manners and modalities in which the world has come to perception, speech and cognition; and the complex cultural histories of thought, practice and learning.

Universities were for a very long time anchored by the page and the book as the material support for learning and protected by the power of religious or political institutions. It was held in place by the man of knowledge called the 'professor', who as one-who-professes transmitted knowledge to the pupils with a messaging system that primarily moved in only one direction. The circuitry was short, with a tight loop. With the emergence of new forms of digital and social interactivity, however, knowledge is now distributed differently.

> The nodal point of this epochal disequilibrium is found in the materiality of the new support of knowledge, the computer, in which the instruments and figures of knowledge are unified (*TS*, ix). This precise support conditions, in a determining manner, the figures of the new knowledge, in the context of an influence of information on knowledge that is initiated by writing and continued by printing. This key to the material history of the supports of knowledge illuminates the construction of Western reason on its physical bases and permits the grasping of the bond between information technology and the new universality of knowledge (*TS*, x). (Polizzi 2000, 264)

We know that the 'homogeneity' of the new digital space is not evenly and fairly distributed – place, resources and identities continue to produce eddies, impasses and inequalities within the materialized ethereal space of communicative signals – but Serres is correct to gesture towards the sense that something fundamental is shifting both beneath our feet in the foundations of knowledge and above us in the air swarming with signals. Shaping, flaking, incising, carving, painting, writing, printing, photographing, filming and, now, cloning and coding.

Thumbelina (alongside Tom Thumb) is the figure that Serres employs to represent this transformational shift in culture, education, identity and the configuration of the digitizing body (both politic and personal). Thumbelina – who passes through us all – works in the world not by deduction or induction (although these continue to be activated along the circuits) but, instead, through what Serres (2015, 71) calls an *algorithmic* mode of thought that includes 'rules of order, sequences of gestures, series of formalities'. Thumbelina texts a friend and there is an algorithmic procedure at work that has an efficacious edge that does work in the world.

Gilles Deleuze (1990, 155) has formulated an analogous series of exchanges by his insistence that the 'real difference is not between the inside and the outside, for the crack is neither internal nor external but is rather at the frontier. It is imperceptible, incorporeal, and ideational. With what happens inside and outside, it has complex relations of interference and interfacing of syncopated junctions – a pattern of corresponding beats over two different rhythms'. Interfaces of interference, meaning and implications, self-learning codes that branch out from decision-trees and the unexpected as it emerges from all the established policy, procedures and programmes of an institution are the multifaceted languages of the university of invention. This is the auto-poetic machine of the university that must re-create itself along the margins and on the fly as it combines the most rigorous of planning with an expectation of the unexpected. This is art, this is science, this is angelic play.

Serres (1995b) underlines not only the interface, but also the fuzziness, complexity and creativity that plays along such edges. 'Have you noticed the popularity among scientists of the word *interface* – which supposes that the junction between two sciences or two concepts is perfectly under control?' (emphasis in the original) he asks Bruno Latour.

> On the contrary, I believe that these spaces *between* are more complicated than one thinks. This is why I have compared them to the Northwest Passage . . . with shores, islands, and fractal ice floes. Between the hard sciences and the so-called human sciences the passage resembles a jagged shore, sprinkled with ice, and variable . . . It's more fractal than simple. Less a juncture under control than an adventure to be had. (70)

Serres' itinerary is never a linear progression, but, instead, follows the contours of fractals and passages that cut across epistemic and earthly geographies in an unforeseeable manner that joins, differentiates and sorts in ways that the specialized disciplining of departments can neither predict nor control.

Such disciplinings, which are woven into the histories of modernity, do not fill a pre-established epistemological space, but create epistemic, social and ethical space as they move. What ecstatic structures of the university to come might be able to perform magic, enable through constraints and liberate through windows, doors, stairwells, elevators and skylights? Through screens and virtual realities? How will the forced analogies of the Surrealists be performed in the university offices of senior management, estates planning, human resources and the student union? In another, more common lexicon, these are the sites where paradigm shifts are occurring. One word for this capacity of fundamental innovation, a problematic one to be sure, is 'intuition', for innovation is 'impossible without that dazzling, obscure, and hard-to-define emotion called intuition. Intuition is, of all things in the world, the rarest, but most equally distributed among inventors – be they artists or scientists' (Serres 1995b, 99). Intuition is equally partitioned across all parts of the university and all forms of inquiry; it is both dazzling and obscure, in the usual non-binary (re)combinations of Serresian logic; and it carries with it affect, hybridities and quasi-objects rather than only the presumed logical purity of reasoning (although it always ends up linking with these operations).

The betweening of the interface where Thumbelina stands, device in hand, is not the pre-established juncture of a railway shunting yard, of a controlled circuit board, or of a regulated street crossing, but is, instead, the crossing point of an adventure. Where, and how, will she set off? What will cause her the greatest agony and the most surprise? How, if at all, will she step out of the scenario programmed by the world? How will she learn to learn? *Relationality*: the interface which Thumbelina *is* . . . and style is the sign of an interface. This is a *transversal* or *vectorial* movement of thought. 'There is neither beginning nor end; there is a sort of vector . . . Vector: vehicle, sense, direction, the trajectory of time, the index of movement or of transformation. Thus, each gesture is different, obviously' (Serres 1995b, 104). Serres makes gestures of thought as a style of transversal writing that responds differently to particular territories of questions. Algorhythms provide something like a compass, useful for orientations and for setting-off.

In relation to the powers of science and the new power of the digital, Serres would claim both that we humans dominate the earth – we are as powerful as tectonic shifts – but also that, in turn, we are played by the movements of the game of what we can still, if just barely, call 'nature': we attempt to follow the infinite speed of the flight of the ball of that quasi-object called 'knowledge', which has no master. In order to indicate the prepositional and angelic character of all of these terms, Serres coins the terms 'quasi-object' and 'quasi-subject'. 'The quasi-object is a form of mediation which originally comes into being as a way of fixing or stabilising social conflicts which might otherwise tend to degenerate into absolute chaos, or all-out, all-against-all war. In that it marks the boundary between the subjective and the objective' (Connor, *Milieux*). With Serres' invention of the operations of the *quasi-*, which is relational and not substantial we are 'dealing here with something very different [than the general formalities of deductive logic] – that is, taking seriously the particularities of sites, the unpredictability of circumstances, the uneven patterns of the landscape and the hazardous nature of its becoming. In short, again: how to think the local? Which means: is there a science of the particular?' (Hénaff 1997, 72).

The dominant model of the declarative history of reason has actually long been complemented by a 'more humble and more subservient reason, which calculates, organizes, adjusts and goes from local model to circumstantial solution . . . From now on, [reason] can offer another figure: *what is flexible, woven and circumstantial* . . . Henceforth, it is possible to elaborate rigorously a transition from the local to the global by algorithmic procedures' (Hénaff 1997, 73; my emphasis). Vectorial thought takes specificity along for the ride and lends an ear to circumstance. Like an arch-angel, which generates the local wherever it goes without being bound to the local, it flies close to the sea and then, in less than an instant, beyond the opening singularity.

Although no analysis can per se be transferred to another conjunctive series of events, problems or questions – singular events demand singular analyses – we can learn from Serres the inclinations of perpetual translation that will teach us to be more attentive to what might be occurring in the neighbourhood as we interweave the unexpected encounters together, under and over, we might come to know, or even be, something different. This is perhaps the primary task of the concept of the emerging transdisciplinary polyversity in which the *uni-* is just one form of the local. This is angelic invention.

Angels do not know what they will inspire as they riffle the water and gust about in the trees, the grass, galactic nebulae or in the hair of Venus on her half-shell. But when they passed by the home of Michel Serres they inspired a world consisting of dynamic sets of infinite worlds that includes a new image of the university. As we have insisted, there can never be a stabilized form of 'the' university, for Serres is always attuning his work to multiplicity and relations, not unified definitions and essences. The university, as a concept, has always since its inception – whenever and wherever we might assign that inception – included within itself a utopian momentum, but momentum must stay on the move. It is not an analytic concept, one in which its predicates are included in it as subject, but a wide-open synthetic experiment with the myriad constellations of the world in which we find ourselves.

Invention, creativity in all of its forms and along all possible trajectories will be its raison d'être, projecting the possibilities for make-believe, for concoctions and contraptions and for provisional communities of learning that come and go, that set up shop and then strike the set. The buildings – for there will still be buildings, since humans, however nomadically, dwell always architecturally upon the face of the earth – will be flexible spaces and the art studios, library nooks, tablet cafés, digital and chemistry laboratories will be interconnected in new ways, both physically and virtually. There will be a poetics of resonances at work across which will play the games of passing, relaying and the rhythms of the jazz that always accompanies knowledge. (This is why there are now the behaviours of surfing, swiping and clicking.) There will be relations and relations of relations. It will be dizzying, dazzling. Modular and plugged-in, plugged-out; self-paced learning with groups of fellow travellers; projects, new distributions of space and time, exploring the proximities that most demand attention and create the greatest buzz. A different humming will mark the university of invention.

'We never know in advance', Deleuze (1995, 165) reminds us, 'how someone will learn: by means of what loves someone becomes good at Latin, what encounters make them a philosopher, or in what dictionaries they learn to think.' We do know, however, that through loves, unexpected encounters and various dictionaries learning does occurs and we have always begun to take on the tasks of transmitting the conditions for learning to others. The University of Angelic Invention will be radically reorganized as a series not of discrete blocks of disciplines and schools, but tranversally, vectorially, algorhythmically.

All the spaces, with their times, of transformation 'emerge from this hodge-podge of abundance whose merit lies in having taken diagonally, askew, crosswise, many of the usual and stupid distinctions of philosophy. Translation is both a praxis and theory; turbulence is a stable and unstable phenomenon where liquid moves and stays in randomly fixed form; the organism – by body – is now an exchanger of time' (Serres 2007, 72). The university that is tentatively showing itself will engage with temporalities and condensations, with zigzags and zones, with the weather approaching from the future. It will require impossible amounts of learning, but we have each other, we have the machines and we have the cosmos as companions.

Following Kant and many others, Serres (1995b, 184) observes that all that we have is 'education to make us adaptably prepared for the future'.[4] All we have is learning to learn. As a species that is augmenting natural and random selection, we need to learn, quickly, how to do things differently, to teach and learn with much greater imaginative flair and a sense of a vectoral *phronēsis* that attends to the call of the other, whether human, oceanic, organismic, ideational, affective, mechanical, micro or macro. Learning is mutability towards the expansive, resonant and playful. This has become an ethical imperative: create joy.

The University of Angelic Invention will take place simultaneously here-and-there; it will be transversely folded across all knowledge domains of the past and future; it will welcome the curiosity of all comers, of those who arrive at the door, the threshold; it will have passageways and pathways, spots of focus and moments of intensity; it will be a field of jousts and three-dimensional games of chess, but without the cutting of the blade or of bloodshed; it will speak with the glorious flutters of wind on the water and it will have to help us all invent a new language for all of this.

The angels, incessantly messaging with the invisibility of their wings and fingers, give us 'a philosophy of communication, traversed by system of networks and interference and demanding, in order to be able to establish itself, a theory of the multiplicities, of the chaos, hubbub and noise that come before all theory' (Serres 1993, 93). Noise-information-noise: we are relearning to listen.

> Every work of art or science, however large or small, consists in catching the wave just right, and following it all the way down the line, for as long as possible, riding the crest, surfing, until we come to the inevitable final fall. If inspiration is in short supply, we fall straight away, or don't even get moving in the first place; but a masterpiece travels fast, moving but immobile, in a long horizontal plane, just slightly off-balance, on invisible lines of force that are etched imperceptibly on the wall of water. (34)

Learning the beginnings of how to write an almost visible and always disappearing line on an indomitable wall of water – and we are all off-balance as we write – is how we will create a different university. This is inspiration, in- and out-spiralling. This, and not utopia, is what the angels sing *about* – there is a referential function to messages, but that is only one, and often not the most essential, of messaging's many functions – and what they, in singing, invite us to bring into being.

The dappled university of the *trans-* cannot be pinned down to a specific past, present or future and will never consist of a clear and distinct 'idea of the university', since that which shines in its own brightness also creates shadows, chiaroscuro and opacity in the act of shining. Ideas are obscure, but they vibrate, glow. It will be not a unity or a synthesis without glitches, but a '*syrrhèse*, a confluence, not a system, a mobile confluence of fluxes' (Serres 1995b, 122). The procedures, policies, curricula, people, screens, labs, studios, media stations, architectures, codes and particle accelerators of this university of difference will not, therefore, fit perfectly together into a whole. There will still be strife across and within the faculties and in the Senate Chambers bedecked with the University Crest, for the *polemos* is always active. Robes, with their traditional velvet trim, tassels and golden resplendence are always already the motley costume of the Harlequin.

If the university recognizes the opening of this invitation, it might act to create a new platform that sings the earth, the clouds, and the voyager headed into interstellar space incised with our bodies, our signals, our conceptions, our maps and Johnny B. Goode. It will learn from the distant proximities of the digital and the incalculable differentials that keep it all moving in its momentum so that it can better play its role as a quasi-object and a quasi-subject atop a thin interface along the white-tipped waves of gravity arriving from the unimaginably distant past as they curl in and pass through all that is in the offing from all the other openings. The angels will rock and roll as we set out to invent a polyphonic university that burns, with a slow and fierce velocity, with an enigmatic joy.

References

Allen, Graham (2016). The Transparent University: Kant, Derrida, and a New University Law, In: John W. P. Phillips (ed.), *Derrida Now: Current Perspectives in Derrida Studies.* Cambridge: Polity.

Brown, Steven D. (2002). Michel Serres: Science, Translation, and the Logic of the Parasite. *Theory, Culture & Society* 19.3: 1–27.

Connor, Steven. Michel Serres's Milieux: http://stevenconnor.com/milieux.html (15 April 2018).

Deleuze, Gilles (1990). *The Logic of Sense.* Trans. Mark Lester with Charles Stivale. Ed. Constantin V. Boundas. New York: Columbia University Press.

Deleuze, Gilles (1995). *Difference & Repetition.* Trans. Paul R. Patton. New York: Columbia University Press.

Derrida, Jacques (2016). *Heidegger: The Question of Being and History.* Trans. Geoff Bennington. Chicago: University of Chicago Press.

Diez, José A. (1997). A Hundred Years of Numbers: An Historical Introduction to Measurement Theory 1887–1990. *Studies in the Philosophical History of Science* 28.1: 167–85.

Hénaff, Marcel. (1997). Of Stones, Angels, and Humans: Michel Serres and the Global City. *SubStance* 83: 59–80.

Kant, Immanuel. (1900/1803). *On Education.* Trans. Annette Churton. Boston: DC Heath & Co.

Kant, Immanuel. (1990). *Kant on Education.* Trans. Annette Churton. Boston: D.C. Heath and Co.: http://oll.libertyfund.org/titles/35.

Kunzru, Hari. (2008). Interview with Michel Serres. Blog Post, Wednesday, December 17: http://www.harikunzru.com/art-and-music/michel-serres-interview-1995.

Lyotard, Jean-Fraincois (1984). *The Postmodern Condition*. Trans. Geoff Bennington and Brian Massumi. Minneapolis: University of Minnesota Press.

Polizzi, Gaspare. (2000). Hermetism, Messages, and Angels, Trans. Trina Marmarelli. *Configurations* 8.2 (Spring): 245–70.

Poovey, Mary. (1998). *A History of the Modern Fact: Problems of Knowledge in the Sciences of Wealth and Society*. Chicago: University of Chicago Press.

Rendell, Jane with Pamela Wells. (2001). The Place of Prepositions: A Space Inhabited by Angels. In: Jonathan Hill (ed.), *Architecture: The Subject Is Matter*. London: Routledge.

Serres, Michel (1993). *Angels: A Modern Myth*. Trans. Francis Cowper. Paris: Flammarion.

Serres, Michel (with Bruno Latour). (1995). *Conversations on Science, Culture, and Time*. Trans. Roxanne Lapidus. Ann Arbor: University of Michigan Press.

Serres, Michel (2007). *The Parasite*. Trans. Lawrence R. Schehr. Minneapolis: University of Minnesota Press.

Serres, Michel (2015). *Thumbelina: The Culture and Technology of Millennials*. Trans. Daniel W. Smith. London: Rowman & Littlefield.

Shore, Cris and Susan Wright. (1999). Audit Culture and Anthropology: Neo-Liberalism in British Higher Education. *The Journal of the Royal Anthropological Institute* 5.4 (December): 557–75.

Warburg, Aby. (1999). *The Renewal of Pagan Antiquity: Contributions to the Cultural History of the European Renaissance*. Intro. Kurt W. Forster. Trans. David Britt. Getty Research Institute for the History of Arts and Humanities.

8

The world, the mat(t)er of thought

Rick Dolphijn

We should not act and speak like children of our parents.

– HERACLITUS

May '68 against the history of thought

In September 1988, in an interview with the *Magazine Littéraire*, Gilles Deleuze (1995, 155) announced his plans for the near future: 'I want to write a book on "What is Philosophy?" . . . Also, Guattari and I want to get back to our joint work, and produce a sort of philosophy of Nature, now that any distinction between nature and artifice is becoming blurred.' The joint philosophy of Nature was never realized, but *What Is Philosophy?* was published in 1991 (officially authored by Deleuze and Guattari although it seems that (practically) all of it was written by Deleuze due to Guattari's depression). In *What Is Philosophy?*, however, in chapter 4 entitled 'Geophilosophy', the philosophy of Nature announced in the aforementioned interview does seem to become apparent. Very much writing 'with, in and according to' nature, and not so much 'about it', 'Geophilosophy' conceptualizes nature as Deleuze had – only fragmentarily – done throughout his career, from his earliest courses on Jean-Jacques Rousseau to the infamous 'Geology of Morals' (written together with Guattari) in *A Thousand Plateaus*. Meanwhile, Guattari himself, supported by Paul Virilio, was working on his *The Three Ecologies* alongside several other shorter texts in which he worked on nature from an activist position (I am talking about the end of the 1980s here, the time Deleuze gave the abovementioned interview). Rethinking ecology not so much as something that 'happens' to the world outside of us, but rather as the set of relations between the social, the mental and the environmental, Guattari did what he was good at: writing political manifestos that captured the momentum, doing philosophy in the proper Marxist manner; not with the aim to interpret the world but to change it. Thus, in *The Three Ecologies*, Guattari (2008) showed how the various crises that dominated the world (at the end of the 1980s) could not be seen as separate from one another: the environmental crisis was in the end no different from the crisis of communities and the crisis of media culture. An ecophilosophical perspective, one which sees that 'the questions of racism, of phallocentrism, of the disastrous legacy of the self-congratulatory "modern" town planning, of an artistic creation liberated from the market system, of an education system able to appoint its own social mediators, etc.' (23) is needed in order to improve the conditions for life on earth.

It was this urgency, this augury, that marked Guattari in every way, and that attracted Deleuze to him from the beginning. Guattari was always the militant, the tireless intellectual (without finishing his university education) whose countless activist projects made him quite a different persona to the established philosopher that Deleuze was already when they were introduced to one another during the student revolts of '68. Summarizing the period before he met Guattari, Deleuze (1995, 13) once stated: 'I was working solely with concepts, rather timidly in fact.' Employed at departments that focused on the history of philosophy, Deleuze, at the start of his career, was writing as the 'traditional metaphysician' he considered himself to be throughout his life. Deleuze was interested in rewriting the History of Philosophy (with capitals) and then in particular the way this History (in France, in the 1950s and early 1960s) had organized philosophy in such a way that it had marginalized those 'modern' philosophers that did not fit the Cartesian/Kantian hegemony. Or more aptly: interested in those philosophers that did not fit the explicit or implicit neo-Kantian dualisms (as expressed by Hegel, Edmund Husserl and Martin Heidegger), Deleuze set himself to rereading philosophers like Baruch de Spinoza, Friedrich Nietzsche and Henri Bergson, as they all questioned these dualisms in their own way.

Events in 1968 changed Deleuze's life for the good and for the bad: he became terminally ill (and had to live on one lung); as a direct result of the student protest a new experimental institute that became the University of Paris 8 (first in Vincennes, later in St. Denis) was erected and became his intellectual home for the rest of his life; he met Guattari with whom he found a way to give a voice to the political urgencies that followed 1968 (the 1970s). And even in the 1980s and after, when Guattari falls into his depressions and Deleuze again starts writing more on his own again, it was this strong involvement in resistance that made him write extensively about the persistence of art in the revolt that was so central to it. Deleuze has always considered himself a child of May 1968 as it was then that working with concepts, as he had practiced this in early writings of the History of Philosophy, turned into creating concepts.

In his introduction to *Anti Oedipus*, Deleuze and Guattari's first joint work, Michel Foucault commented on the unusual partnership of Deleuze and Guattari and on the materialist philosophy these two thinkers produced. He famously stated that the grandeur of this book was not that their thinking should not be considered in line with the History of Philosophy, in the sense that this was not a flashy Hegel. What Deleuze and Guattari (1984, xiii) had produced was rather, as he called it, '*an Introduction to the Non-Fascist Life*' (italics in original). In every way a product of '68, it was with Anti-Oedipus (a philosophical bestseller) that Deleuze found his *responsibility* in philosophy, both in content and in style. A new 'type' of philosophy, of thinking, that refused to accept the rational (white, male, civilized) subject as the starting point of and the paradigm for philosophy. Surely, '68 had changed things dramatically, and this new world demanded a new form of thinking.

August '45 against the history of thought

Michel Serres has a very different relation to the uproars of '68 than most of the talented French philosophers of his generation. Working closely with Foucault (in Clermont-Ferrand and later in Vincennes, where Serres preceded Deleuze as a professor in philosophy), Serres was involved with this generation from the beginning. His major breakthrough, his PhD thesis on Gottfried Wilhelm Leibniz as published in 1968, is a reading of Leibniz against his interpreters, or against the way his interpreters had organized Leibnizian thinking according to the different paradigms of the university (its faculties and departments, its ideas on research, on the philosophical method, on epistemology in general). Through Leibniz, Serres offers us an idea of Leibnizian philosophy that refuses to follow those Leibnizians that perform a disciplinary analysis of *parts* of Leibniz's work, focusing either on differential calculus or on topology, on law or on history, uninterested in the many connections Leibniz himself drew between these fields of study (i.e. disrespecting the fact that the resonances between all of these areas of thought are actually the starting point of Leibniz's thinking).

Serres himself was highly sensitive to these connections and explains this, to give a very basic example, by questioning the idea of space (and time) occurring as a linear sequence (Serres 1968, 286). This idea, as it dominates our idea of space (and time) today (or at least in 1968, when the study was published), was given to us by Leonhard Euler, and it does not match with Leibniz's model of space (or time). Of course, many interpreters of Leibniz understand this when they analyse Leibniz's mathematics, but immediately forget this (*id est*, they implicitly accept Euler's method) when analysing his ideas of law or of history. Reading Leibniz's mathematics means that we start thinking from the creation of an endless amount of molecular relations, a multiplicity of connections or information flows. But of course; the poly-tics of mathematics *is* the poly-tics of law as it *is* the poly-tics of history. Later in his career, in the introduction to *Genesis*, Serres (1995a, 4) refers to his materialism, after Leibniz's never written second monadology, as a monadology of *vincula*, of connections. Thinking the multiplicity, very different from the unity but also from the plural, this thinking of connections is what Serres practiced throughout his career. Whether he writes on pure noise, the cloud, the event, it is *necessarily* a monadology of vincula that his writings give rise to.

It is important to note that, especially in this early phase of their careers, both Serres and Deleuze were not working on a 'grand theory', or on a 'grand revision' of philosophy. Both were not so much *critiquing* the History of Philosophy, as affirmatively mapping the philosophers, artists and scientists close to them; Serres critiqued Descartes/Cartesianism only *implicitly*, Deleuze critiqued Kant/Kantianism only *implicitly*. They never brought these implicit critiques together into a grand Critique. (Deleuze even wrote a small book in which he affirmatively summarized the philosophy of Kant, calling it an 'homage to an enemy'.) Only in the post '68 interviews from the 1970s and the 1980s, when he was writing with Guattari, does Deleuze look back at his own writings from the early days as an attempt to write 'a minor history of thought', a type of non-dualist thinking that he sees, in different ways, at work in the ideas of David Hume, Spinoza, Nietzsche and Bergson. But also with Deleuze, stressing the *minor* nature of this history of thought, this tradition is in no way linear, it is not structured (knows no hierarchy) and does not account for any kind of development (beyond mere change).

With Serres, even more so, his rewriting of the Leibnizian thinking is a praxis *immediately*. His five *Hermès* books, which he started before his Leibniz volume, and which he worked on throughout the 1970s, prove that: from his earliest writings Serres offers us a myriad of connections between science and culture, connections between all the realms of thought, connections that have been there all along, but that we, because of our disciplinary blindness, did not see. An analysis of Leibniz, for Serres, is thus not an analysis of the works of Leibniz and its reception. On the contrary, *Le Système de Leibniz et ses modèles mathématiques* (1968), is at once a philosophy of mathematics and a philosophy of communication. One even wonders if his recent bestseller *Petite poucette* is any 'different' from his first writings in that aspect. Throughout this booklet Serres (2012b, 75) stresses the new type of thinking that surfaces today and only in the end he admits that this 'new' type of thinking, as he develops it throughout the book, was already anticipated by Leibniz (and Blaise Pascal, another one of his heroes). Has he not then given us his reading of Leibniz's philosophy of information, of communication, of mathematics, *Anno Domini* 2012?

Clearly it was not '68 that gave birth to Serres as a philosopher, as it did in the case of Deleuze (and Guattari). For Serres, the momentum that caused his thinking, contrary to Deleuze whose intellectual birth happened during his career, happened long before he started writing. For Serres, it was the dropping of the atomic bomb on Hiroshima, which he 'witnessed' as a teenager, that marked his thinking, as it marked the moment where many of the dualisms that were so central to post-Cartesian thinking were all of a sudden laid open. The Horror of Hiroshima revealed to him, and to many of his contemporaries, how physics was alienated from nature, how epistemology was alienated from ethics and how man was alienated from the world. This threefold trauma drove the young Michel Serres (he was 14 years old when Little Boy was dropped on Hiroshima) to embark on a materialist philosophy of science aimed at mapping the dualisms of Modernity, preferably by showing their mischievous superficiality, together with the many layers of truth that it has hidden from us. As he explains in the interview in this book, the switch that he made from mathematics to philosophy was caused by Hiroshima, just as the rest of his (ongoing) intellectual life was caused by it, from the beginning.

For Serres, '45 changed things dramatically, giving rise to a new world that demanded a new form of thinking. The experimental institute at Vincennes (Centre Universitaire Experimental), as it opened its doors in December 1968 was not then the intellectual backbone of Serres' career (as it turned out to be so for Deleuze). Soon after its start, Serres found out that he was not fond of the experiment that turned out to be the University of Vincennes/St. Denis Paris VIII. It marked the beginning of his academic career, in which he experienced considerable difficulty finding an intellectual 'home' for his unconventional approach. Starting from the break with Canguilhem perhaps (after his 1968 defence), his (post '45) emphasis on the resonances between the sciences and the humanities left him wandering through the various faculties of university (primarily History and Languages departments, see, for instance, the introduction to *Rome*), yet he was never appointed by a Philosophy Faculty. Keeping in mind that Deleuze, who stayed in Vincennes until his retirement, also never got a good position at a high-profile university, one wonders why these philosophers, who were actually *doing* philosophy from the start (not surrendering to the History of Philosophy as so many of the philosophy faculty did), are not recognized by the Philosophy Faculties of the major universities in France (and in the world) for their efforts. Perhaps this is the universal tragedy of philosophy (or at least one of them): more so than in any other area of thought, the revolutionary thinker is seldom recognized by his powerful peers in academia. Let's not forget that even Spinoza and Leibniz, the historical predecessors of respectively Deleuze and Serres, did not get due recognition from philosophy during their lifetimes.

It is important to note that, much more so than Deleuze, Serres started his career from the position of an outsider. Coming not from Paris but from the Aquitaine, in the southwest of France, going to a naval academy studying mathematics and physics, his relation to philosophy grew from the philosophy of science, that part of philosophy so often considered 'the least philosophical' as it so often seems to be too restricted to the (positivist) chronologies of the sciences. With a great interest in those thinkers that combined their interest in the sciences with philosophy (like Leibniz, but also Galileo Galilei and Pascal) from the start of his career, Serres' outsider perspective allowed him to *do philosophy* by reading turbulence theory with Lucretius' poetry and thermodynamics with Émile Zola's novels. As mentioned above, Serres, from the start, refuses to accept the fortresses of academia that split and divide, traversing and rewriting all of these fields in content and style. Not interested in the technical talk that implicitly insists on maintaining the fortress of one's discipline, this is then what doing philosophy is all about. Inspired by all sorts of writers inside and outside of academia (remember his famous claim, often repeated, that the difference between philosophy and literature is produced by university), he was always already working on a monadology of vincula. Neither through explaining (*erklären*) nor through understanding (*verstehen*), this *procedural* philosophy practices a kind of thinking that was traditionally only found at the margins of academia: with medicine and law, within a few areas of mathematics (arithmetic, differential calculus and topology), and perhaps with the way musicology was practiced before its crisis.

A procedural philosophy of sameness

It is for this reason that Serres has always been intrigued by medicine and law/ jurisprudence; for here we see two well-developed parts of academia in which procedural philosophy had always already been practiced. Serres sees a similar non-representationalist and non-dualist approach in some parts of mathematics (algorithmics, topology) as it moves away from the deductive principles of mathematics as well as from the inductive principles of experimentation. This is a much more inclusive, pragmatic and materialist way of thinking compared to how the alpha's and beta's traditionally (at least since René Descartes) further their study of the world. Procedural thinking does not allow for taking up a position 'in relation to' the world (it is deeply non-dualist): it necessarily happens with it, it moves with its movements, changes with its changes. It doesn't reflect upon something, it does something, *it is communication itself, it is the earth itself.* It is the wave of information, as it forms both the surface and undercurrent of the real, the conditions for philosophy to start.

For Deleuze, and many of his Parisian peers, the student protests of May '68 in particular caused them to start philosophy from 'difference'. Jacques Derrida, for instance, dedicated much of his thinking to the linguistic, societal and psychological differences that could not be named, focusing on the margins of thought and not on the centre. Michel Foucault, in *The Order of Things*, stresses that instead of the history of sameness (which is he develops in that particular book), the intellectual focus throughout his career will be the history of otherness. But also in French feminism, in critical studies of race and in all the new developments in Marxism, May '68 serves as the foundation for a contemporary philosophy of difference. (Michel Foucault, in this respect, is in the fullest sense a philosopher who made history into a philosophy of the contemporary – contrary to his own belief). It is crucial to stress here that this emphasis on difference was without a doubt one of the most fruitful revolutions in philosophy as it fed an entire generation of thinkers with ammunition to attack the dominant patriarchal, occidental normalities of that day. The emphasis on difference was a necessity for an innumerable variety of emancipatory processes.

Without delving into the matter too much – many books have already been written on this topic – it is fair to conclude that Deleuze's approach to difference in the end turns out to be much more radical compared to the way many of his contemporaries worked with the concept. Rereading Nietzsche's difference as an *absolute* and not a *relative* difference, preferring to think 'wholly otherness' (as he puts it throughout his *The Logic of Sense*) and not Otherness as opposed to sameness (as Foucault, for instance, proposes this), Deleuze is *not* searching for the recognition and emancipation of the unheard, the unseen and the unknown, like so many of his contemporaries. Emphasizing difference-in-itself, Deleuze instead proposes to raise *another* world (with Nietzsche again), a new kind of people that cannot be considered relatively 'different' from the kind of humanity that dominates our planet and our thinking today. He proposes, as Rosi Braidotti often puts it, a philosophy of freedom and not one of emancipation.

Being a child of August '45, Serres never engaged with the concept of difference in the first place. He begins his book on the origins of geometry by stating that 'our discourses, for at least a half-century, have rumbled with our differences' (2017, x). The grand narratives of post '68 'studies' depend on difference in order to start up their theories of emancipation: 'But by some perverse paradox, difference ends up imposing itself in turn as a universal dogma that everywhere and always forbids speaking forever and everywhere' (idem.). Difference opposes, difference represents, difference legitimizes dualisms . . . One wonders why a *wholly* new earth, a *wholly* new people, would actually need difference.

Here's another (related) paradox to think through. One might think that the importance of Hiroshima in Serres' work comes with a kind of negativity whereas May '68, in the work of others, comes with a kind of positivity. Dropping atomic bombs is hard to consider a joyful event whereas a revolution that aims to get rid of all hierarchies, probably (in a way) is. Its lives in thought, however, work the other way around. For whereas May '68 revealed the many differences that mark society, that split it up and organize its hierarchy, Hiroshima showed us that all the differences to which we implicitly plead allegiance (physics versus nature, epistemology versus ethics and man versus the world) in fact point to nothing else but our short-sightedness. Whereas May '68 can be considered the birth of contemporary critical theory, Hiroshima, as Serres reads it, offers us theories of affirmation without a limit. Hiroshima, in that sense, is the permanent goodbye to Humanism, and to all of its Cartesian and Kantian dualisms that still dominate our thinking. In line with that it is also a goodbye to the way science today is (still) firmly based on a set of modern principles. Isabelle Stengers (2010, 91) agrees with this idea that physics is paralyzed by its understanding of 'physical laws', by 'discoveries', by 'states of things'. Scientific reasoning, she claims, all to often depends on 'factishes' (after Latour); 'fabricated beings that are *as such* capable of an autonomous mode of existence'. Serres (1995b, 84; italics in original) sees this sense of 'ownership' as symptomatic for the whole project of modern science since Galileo Galilei, who he critiques for being 'the first to put a fence around the terrain of nature, take it into his head and say, "this belongs to science", and find people simple enough to believe that this is of no consequence for man-made laws and civil societies, closed in on human relations as they are'. Serres thus moves away from the fundaments of modern science, showing us, as Serres' student Bruno Latour phrased it (after him), that 'we have never been modern'.

Perhaps it is because of his refusal to accept Modernity and its consequences as his intellectual heritage, that Serres, long before the other members of this talented generation of French thinkers of difference (and this includes Deleuze), was able to develop, already in 1990, a type of non-dualist naturalism that is now very much explored in new materialism, continental naturalism, eco-posthumanism. Like Gregory Bateson (see 1972), perhaps the other great maverick thinker in twentieth-century thought, Serres' ability to think transversally naturally moved him away from the all-too-human theorizations that marked the second half of twentieth century thought. Transversal connections of that travel every realm of thought, showed him what the world is saying and allowed him, as it did Gregory Bateson, to *include the earth in thought*. Earlier I quoted Guattari's activist manifesto *The Three Ecologies*, published in 1989, which was directly inspired by Bateson and showed us the transversal connections between the mental, social and environmental ecologies that make up our world (and us in it). Serres' *Natural Contract*, published around the same time, anticipated our current interest in the

Anthropocene and the ecological crisis even more than Guattari (2008, 4), when concluding: 'At stake is the earth in its totality, and humanity, collectively. Global history enters nature; global nature enters history: this is something utterly new in philosophy.'

The mat(t)er of thought

The earth in its totality has to be thought (and produced!) as a philosophy of the contemporary. And then the contemporary should not just be read as the immanence of time, as space is equally fundamental to it (note that in French, extensiveness [*entendu*] is etymologically linked to times [*temps*]). Philosophy does not start with an 'I think', with an independently existing subjective point of view from which the world (as a Divine Readymade) must be observed (as Descartes put it). Philosophy, for Serres, is the immanent mapping of new surfaces (and undercurrents), through resonances and transversal patterns of sameness, as they expand in a multiplicity of actual and virtual directionalities and dimensionalities. This is the revolution in philosophy that Michel Serres practices, since day one (as he couldn't and wouldn't do it *differently*).

His plea for a Natural Contract, which confines humanity to the ongoing exploration of the contemporary, shows this perfectly. Preceded by other studies, for instance *The Birth of Physics*, published in 1977, in which the naturalism of Lucretius (and Archimedes) is given its first rigorous study (often implicitly critiquing Newton and Descartes), he stresses that our day and age is not asking for a Hobbesian social contract and is surely done with the 'Laws of Nature' (that we invented). We need a non-representationalist, monist, materialist contract based on the subtle system of constrains and freedoms that give form to the living earth (*Biogea*) as a whole. Serres (1995b, 48) explains:

> Through exclusively social contracts we have abandoned the bond that connects us to the world, the one that binds the time passing and flowing to the weather outside, the bond that relates the social sciences to the sciences of the universe, history to geography, law to nature, politics to physics, the bond that allows our language to communicate with mute, passive, obscure things- things that, because of our excesses, are recovering voice, presence, activity, light.

For Serres, the task of philosophy is to produce *all* of these bonds while writing/thinking. Thus, he shows us that the social scientist *is* the policeman (as their aim is control), that nature *is* culture's hell (as it undoes its humanism) and that the history of science *is* the history of religion (as they are marked by conflict and tribunals that aim to regulate it). But also, a story about the joy of camping *is* the claim that the archaic and the primitive accompany us step by step; the story about a religious mother loved by her daughter, reborn through her daughter, *is* the story about the cut down tree in China being sung to by the woodcutters, before the trunk puts new roots in their bodies (the etymology of matter is key here . . .).

Drawing these affective patterns of sameness as they surface throughout the text, Serres shows us how seemingly different (or separate) events belong to the same multiplicity. Especially today, as we are living in the days of the Anthropocene (as scientists refer to it), the days in which the earth has entered history and philosophy again, enforcing (and in search for) new alliances, new resonances and harmonies, we have to open ourselves up to those patterns of sameness; the materialist and earthly patterns have never been so important in philosophy! Latour (2017) recently stressed again that Modern culture was never involved with the earth. Christian Gnosticism simply refused to think about it, capitalism took all the opportunity to exploit it, and political thought was turned into sociology; as Serres (1995b, 3) puts it 'Does anyone ever say *where* the master and the slave fight it out? Our culture abhors the world.'

In more recent work, for instance, in *Biogea*, Serres proposes the setting up of a parliament where the four elements and life (Water, Air, Fire, Earth and Life; WAFEL) decide about the future of the earth together. WAFEL is not based on relations of power, it is not a -cracy or -archy, but reminds us of isonomia, that pre-democratic system of non-rule that was realized in ancient Ionia (where Greece meets the west coast of Turkey). Proposing a radical break with all the systems of kinship and all the systems of power that we are familiar with (the -cracys and -archys; see Karatani 2017), the time has come to think about ways of living together *not* based on power but on the equal right to speak. Only this way can we listen to the voices of the forest, to the algae and the stones and, more importantly, to *all* of the new alliances (think of plastiglomerate) that populate the biofilm that embraces the earth, that makes up the political ecology of Gaia today. Inevitably, as this is a political system based on the ongoing negotiations between everything instead of a fixed system of human laws, WAFEL places humanity at the margins of the earth, where it belongs.

But even better, in WAFEL, instead of constantly claiming this right to speak (grammatica, retorica and dialectica still form the core of our educational system), WAFEL will want us to listen. Or, to be more precise, rather than using all of our intellectual stills to identity and name the social differences that segmentize human society, WAFEL asks us to develop the art of listening; to open our ears (and with that, all of our senses) to all of the sounds/voices we never heard before. WAFEL suggests a parliament that promotes the expression of everything in terms of words, formulas, feelings, but also in forms unknown to us. WAFEL is realized in interfering networks, in resonances, reverberations, fuzzy codes and wavy patters. Human thought is now forced to open itself up to all the other human and non-human voices and non-voices. And as such it will more easily understand itself as one of the many immanent realizations of the swerve.

The multiplicity of negotiations between water, air, fire, earth and life, termed WAFEL, need not necessarily be considered in terms of a *possible* non-human form of isonomia, a *possible* elected parliament that bears responsibility for the entire earth (and not just for humans). In *Biogea* Serres repeatedly refers to it as a virtual parliament, one that has always already been there, that has been real as real can be long before we even set foot on this earth (indeed, that gave birth to us). WAFEL is then not to be seen as an 'alternative' for our current form of government. It shows us the true virtual called *connectivity*. Getting rid of the Modern, dualist noise, that, for instance, defines our present (totally dysfunctional) parliaments, WAFEL always already takes place in all matter, in all clustered bodies that live Biogea. The term WAFEL then points us at nothing less than the decision-making process as it has always already been taking place. Thus, it is also *by all means* a process that takes place 'inside' of ourselves (we too are WAFEL, though we have forgotten). That is what Serres means when he says that contemporary knowledge is the status of Biogea. It is one with it. In *Biogea* (2012a, 36–7), echoing the quotation from Bateson above, Serres shows us how these processes make up for Serres too (temporarily):

> The age-old union of [Garonne's] bed and floods with my veins and arteries reproduced my entire life before lakes and torrents. I gushed forth from her water, from the same water the world was born . . . Nothing could be more stable in memory and history than the processual turbulence that eddies in this vortex, like in my body – that middle-knot of Garonne – . . . In my body and across the world, Garonne circulates. My time goes and life passes, this eddy remains.

Thought and the earth: the nomad and the naval officer

In an interview with Serres just after the tragic death of Deleuze, the interviewer asked Serres (1995c) about his academic and personal relation with Deleuze:

> Deleuze was – I lost my best friend last month, because Deleuze was my best friend. I admired him. I loved him. When we were young we were very separate. Together we invented the term *amis de vieillesse*. You know the expression *amis de jeunesse*? We were not *amis de jeunesse*. We became *amis de vieillesse*. And why? Because we are a little bit brothers. I think that Deleuze is a geographer, and I am too a geographer. We are not historians.

Amis de vieillesse. They could only become friends of old age. In an interview that Deleuze gave to Claire Parnet, known as *l'Abécédaire*, Deleuze talked of the pleasures of old age, which according to him had to do with having been let go, with not being burdened by society. Like Robinson, old age drifts us into a situation in which there were no Others, in which the chronologies of the present are not forced upon us anymore, in which we can perhaps more easily search for a wholly other life. But for Deleuze, this new situation that he finds himself in was not of course 'invented' by his retirement (as the interview seems to suggest). Throughout his writing, for instance in the first appendix of *The Logic of Sense*, the ideal of a world without Others, is being conceptualized (indeed, by reading Michel Tournier's novel *Friday*), or better even: *is being realized*.

What is this world without Others? Rereading Tournier, it is clear immediately that this wholly other world (the island Speranza) refuses to be lived like a world with Others. The systems that structure the world with Others (for instance, capitalism and religion) cannot be carried over into this other world. Missing too many cogwheels these machines just do not work anymore. Robinson therefore has to search for a new form of life. A life in which not just the structuring systems disappear, but, with that, practically all of the qualities that we (implicitly) associated with them. A world without others is thus not 'different' from a world with Others. On the contrary, it is a world in which, after the shock (the explosion) wholly new patterns of sameness have to be mapped out. Of course, these new patterns have little to do with Modernity anymore, nor with the dualist oppositions that came with them. Of course, these new patterns do not follow from a History, nor do they project us a future. These new patterns, instead, are realized in the contemporary. They do not define the here and now, but, in traversing the world without others (or the island) they produce something which probably does not allow itself to be named (if only because language is the instrument of Others), but that in every way makes the contemporary. It can be felt. It definitely resonates in many ways with the new situation. It reverberates with the land and the sea. The sea that has withdrawn from the new land, uncovering its unforeseen chaos.

Both Deleuze and Serres began their careers as historians. Deleuze rewrote Spinoza (and Hume, and Bergson, and Nietzsche). Serres rewrote Leibniz (and Auguste Comte, and Lucretius). Both of them (immersed in the world of Others, as insider or as outsider) had to *discover* that they were *geographers by birth*. They had to release themselves from the burdens of the present, finding out how to explore the contemporary. Thus they found out, through their respective births, that philosophy is about exploring new land as it reveals itself to us. Philosophy is about territorialization, deterritorialization and reterritorialization as Deleuze would put it (remember his claim, with Guattari, that thinking takes place between a territory and the earth).

Throughout his oeuvre Serres practices the idea that thinking is all about this transdisciplinary form of mapping that necessarily practices a materialist, naturalist and non-human study of the earth in its totality. Geographers traverse the here and now, give rise to a new world, as Serres (1995b, 53) stresses by reading the history of this concept: 'Geometry in the Greek manner goes back to the Egyptian *Maat*. This word signifies truth, law, ethics, measure, and portion. The order that comes out of disordered mixture, a certain balance of justness and justice, the smooth rectitude of a plane.' After the flooding (of the Nile river), the *harpedonaptai*, Serres adds, was the one practicing *maat*. The *harpedonaptai* was the one giving new birth to law, setting up a new pact with the world, with nature. It set up a new contract under negotiation that preceded not only law but also science and politics. The *harpedonaptai* is the true philosopher, the true geometer who did not start from the History of thought (or of law) but explored the new world, that is, the world without Others.

But then; as both of them spend their lives exploring their respective births, remembering what made them think, what always already drove them as philosophers, they met. In the end. In the beginning. The friendship of old age between Deleuze and Serres could *only* be established in the contemporary and according to this world without Others. Their brotherhood needed to start from this new world. Of course, this was a world always already there, *ergo*, when talking of 'youth' and 'old age' I am by no means implying a chronology (staying true to Leibnizian mathematics). The friendship between Serres and Deleuze was always already there. And it will stay for eternity. It was circling the present until, in the early 1990s, it found its momentum for realization. A philosophy in friendship. A philosophy of nature, no doubt. Without limits.

Philosophically, even more so than his friendship with Guattari (which was an *amitié de jeunesse*, a revolutionary love from May '68), the friendship between Deleuze and Serres holds a very powerful promise for studying the contemporary. Deleuze and Serres never wrote a text together. But the writings of Serres today, and the many writings by academics who have now somehow started practicing their *shared* ideas, are installing a continental naturalism that was born from the eternal friendship between Deleuze and Serres. Forever, like a statue, this friendship resists and keeps on interfering in the present. Indeed: 'A living body covered with stones' (Serres 2015, 181). It is a space that asks for a thorough exploration, or mapping, in our earthly times.

References

Bateson, Gregory. (1972). *Steps to an Ecology of Mind*. Chicago: University of Chicago Press.

Deleuze, Gilles. (1990). *The Logic of Sense*. Trans. Mark Lester with Charles Stivale. Ed. Constantin V. Boundas. New York: The Athlone Press.

Deleuze, Gilles. (1995). *Negotiations*. Trans. Martin Joughin. New York: Columbia University Press.

Deleuze, Gilles and Felix Guattari. (1984). *Anti-Oedipus: Capitalism and Schizophrenia*. Trans. R. Hurley, M. Seem and H. R. Lane. Minneapolis: University of Minnesota Press.

Deleuze, Gilles and Félix Guattari. (1994). *What Is Philosophy?* Trans. Hugh Tomlinson and Graham Burchill. London and New York: Verso.

Guattari, Félix. (2008). *The Three Ecologies*. Trans. Ian Pindar and Paul Sutton. New York: Continuum.

Karatani, Kōjin. (2017). *Isonomia and the Origins of Philosophy*. Trans. Joseph A. Murphy. Durham: Duke University Press.

Latour, Bruno. (2017). *Facing Gaia*. Trans. Catherine Porter. Cambridge, Medford: Polity Press.

Serres, Michel. (1968). *Le Système de Leibniz et ses modèles mathématiques*. Paris: Presses Universitaires de France.

Serres, Michel. (1995a). *Genesis*. Trans. Geneviève James and James Nielson. Ann Arbor: University of Michigan Press.

Serres, Michel. (1995b). *The Natural Contract*. Trans. Elisabeth MacArthus and Willian Paulson. Ann Arbor: University of Michigan Press.

Serres, Michel. (1995c). Interview with Hari *Kunzru*. Found at: http://www.harikunzru.com/art-and-music/michel-serres-interview-1995.

Serres, Michel. (2012a). *Biogea*. Trans. Randolph Burks. Minneapolis: Univocal.

Serres, Michel. (2012b). *Petite poucette*. Paris: Le pommier.

Serres, Michel. (2015). *Statues*. Trans. Randolph Burks. London: Bloomsbury Academic.

Serres, Michel. (2017). *Geometry*. Trans. Randolph Burks. London: Bloomsbury Academic.

Serres, Michel (with Bruno Latour). (1990). *Conversations on Science, Culture, and Time*. Trans. Roxanne Lapidus. Ann Arbor: University of Michigan Press.

Stengers, Isabelle. (2010). *Cosmopolitics 1*. Trans. Robert Bononno. Minneapolis: University of Minnesota Press.

9

The grace of extinction

Patricia MacCormack

There is a beautiful juxtaposition between a materiality towards which humans within the social contract are indifferent, and confounding spectacles (and spectres) upon which we are dependent in the work of Michel Serres. While many other philosophers have emphasized the need to collapse Cartesian binaries, and post-structuralism would suggest we are in an age which is attempting to do so in order to create a more vitalistic ethical world, Serres reminds us there remains a fissure between the two which is unresolved and which our anthropocentric unconscious loves. In *Times of Crisis* the abstractions come in the form primarily of money but also media created false antagonisms of good and evil, fear and placation (the only form of somnambulistic pleasure left in the social contract). Binaries persist in what Serres (2014, 24) names 'an era called anthropocentric' and yet many of those championed by justice movements and alterity activists are turned around. While other post-structural philosophers claim the endless repetition of signs in capitalism make a finitude of the infinity of the cosmos Serres states that '[b]riefly stated, the unlimited potential of research, progress, or rational and technological exploitation was and still is aimed at a finite whole of concrete, inert or living objects. To use an image, here humanity's infinity faces the world's finitude. Let us remember we believed the contrary' (24).

Serres chooses to seduce with the expectation of 'exploration' and exchanges for 'exploitation' because his work diligently attends to nature by showing that humans are barely capable of apprehending even a vague concept of nature as something of which we are a part or even something to which we could be opposed by culture or the social. Nature is a finite matrix. We do not live in it or as part of it. We parasite off it, we overcode it with the social contract and we defile it with our abstraction of its materiality:

> A living species, ours, is succeeding in excluding all others from its niche, which is now global; how could other species eat or live in that which we cover with filth? If the soiled world is in danger it's the result of our exclusive appropriation of things. So forget the word *environment*, commonly used in this context. It assumes that we humans are at the centre of a system of nature. (Serres 2002, 32; emphasis in the original)

Ethical relations require the capacity to open the others' ability to express an awareness of one's experience of affect from another. This occurs multidimensionally. Concerns for the environment convert the being of all other entities to which an ecological ethics would connect us into a plane of human occupation which humans manage and manipulate to best serve an anthropocentric understanding of these relations. It is entirely monodirectional and appropriates any experience other than a human perception of what that experience could be. Serres is admirably dubious about human capacity for any other kind of apprehension, not least because the drive to apprehend itself is a technique of control exerted by the human, belonging to what he calls the Order of Mars. In a neat twist, the human compulsion to know the world has created an infinity of empty modes of communication, exchange and production circulated between humans that entirely extricates the non-human world from being capable of expression, and even less affect. Taking these concepts into account, this chapter will explore what an 'environmental' activism could look like from a Serresian perspective when we ask how we can open to nature. Ultimately while the answer 'we can't' may seem nihilistic, Serres offers us the concept of grace with which we can gift nature (a gift without demand or reciprocity) and I suggest this act of grace is useful for activism concerning abolition (the liberation of non-human animals from human use) and human extinction.

'If you wish to save your soul, your breath, your voice, venture to lose them' (Serres 1995, 59). The idea of human extinction is increasingly appealing in a variety of philosophies and activisms spanning contemporary nihilistic Lovecraftian theories (such as the OOO) to VHEMT and anti-natalist politics. However it also has appeal for vitalistic material ethics, which values immanence and finitude. The value of immanence far from a hedonistic emphasis on the now slows down the hyper-velocity of post-structuralism which is trans-human or technocentric. It also refuses an atrophied history that read forwards constitutes the definition of present truth or what Serres (1995, 91) critiques as the structuration of life via classification and science which Serres places in opposition to love. Understanding immanence via Serres it is attendance to the natural contract as it is now as a state of turbulence or clinamen between all entities forming a consistency. This both refuses the punctuative violent ruptures that the human perpetrates through both the act of classification (always hierarchical, always with the classifier at the zenith) and the effects of all human acts upon the natural contract from which the human has extricated himself and yet within which he causes particularly aggressive and devastating ruptures. Finitude similarly can be taken from a nihilistic or vitalistic perspective. Post-structuralism has long mourned the death of the human only contingent on it being replaced by something superhuman, while for feminism and other minority politics, attendance on the 'not quite counting as human' subjects in the world today has seemed somewhat of a luxurious enterprise. For ecosophists and abolitionists (those who seek not animal rights as equivalent to human but seek the abolition of all use of non-humans by humans) minoritary politics is not limited to those that serve subjects already classified as thus – women, non-whites, the disabled, LGBT and so on – but those entities who cannot enter into the social contract except as collateral or capital – non-human animals, the environment. Further ecosophists and what I have elsewhere called ahuman philosophers refuse the act of classification at all as visibility drags the minority up to the level of the majoritarian only with their language and on their terms and in practice resists difference, relationality and singular ethical interaction. 'And it is classification, on the contrary, that is negative, it is coding that operates in a negative manner, it is the concept in general, and determination, that is negation' (Serres 1995, 99) and that leads to hatred and destruction (99). Classification as the structuring drive of the social contract attempts to overcode chaos which Serres sees as a natural terrain and a positive force: 'But I say positive chaos. Spinoza does not say otherwise: determination is negation. Indetermination is thus positive, and yet we express it with a negative word. I am simply writing the positive concepts of the under-determined, the undetermined, the positive concepts of the possible' (98). The finitude of the human calls for the end to human classification which already destroys the world before it is encountered and which creates a social order that replaces material actuality with the empty signifiers of exchange and value that lull the human into forgetting materiality, even forgetting the self so that it becomes almost impossible to say we are self-serving anymore. For Serres anthropocentrism has little to do with human flesh. It is this Order of Mars, this aggressive, destructive drive for determinism through social coding that makes us forget the world and unable to access nature. Perversely it also gives the human a sense of infinity in two ways – first because it overvalues the empty signifiers (money, subjectivity, scientific and media representations of truth as constituting 'reality') and these signifiers are adaptable and their tendencies if not their specific signifieds follow us through the ages in different guises but via the same structures and social contract so their eternity makes us forget the finitude of the world. Second because it makes us forget ourselves as fleshly incarnations which do not belong to groups, tribes or species but which are turbulences within the natural contract. By adhering to groups, even those admirable group minority politics for which we continue to fight, we are negating ourselves and all lives by adhering to the seductions of perceived infinity in belonging to defined categories and the many battles they fight, win, lose. We belong to a classified determined group and our own internal multiplicities and disjunctures are lost. If this promise of infinity drives us, even in minority politics, we are negated both in our difference and altogether as singular instances of life (thus so are all other singular instances of life) because our infinity is determined, our possibility is determined. To belong to the social contract is to already belong to the Order of negation, of determinism, of Mars, 'stable, immutable, redundant. It recopies the same writings, with the same atom-letters. The law is the plague. Reason is the fall. The reiterated cause is death. Repetition is redundancy. And identity is death. Everything falls to zero: the null point of information, the emptiness of knowledge, non-existence' (Serres 2018, 134). Serres shows us here two kinds of death. Deterministic knowledge – of subject, of law, of science – creates death in the wake of its destructive war on the disjunctive clinamen that is every singularity of life ('What is a living being? A thing in equilibrium' [147]). It also prevents the equilibrious but always in movement singularity from forging alliances and collisions with other beings. As these relations change the nature of each entity the social contract would see this as the death of the determined object. This death Serres – and I – welcome, because it is the necessary death of the atrophied perceived subject within a classification defined by history and a projection of futurity, and valued through its opposition to other entities and its place within the hierarchy. Serres is explicit about these two kinds of death – the death perpetrated by the social contract where opposition and dividuation wage war in the Order of Mars, and the death of the equilibrium of an entity in the jubilant opening up to the perturbations from other entities which is the creativity of Chaos in nature in the Order of Venus.

Post-structuralism has already celebrated the death of the subject. Read with Serres however there are some issues with the ways in which these have occurred. Replacing the human with many groups of humans fighting for equality retains the law of man which operates two functions Serres sees as destructive. First the act of perception is an act of violence for Serres as it inevitably defines by demarcation and leads to both atrophy and a refusal of immanence. His turn to physics addresses these concerns: 'What once again is physics? It is the science of relations. Of relations in general between atoms of various families . . . Venus assembles the atoms, like the composites. But she is not transcendent like the other gods, she is immanent to this world, she is the being of relation' (Serres 2018, 148). Serres describes relations with the word nature: *foedus*. Nature is the mesh of relations that is positive chaos which is vastly different to natural law, which is a fantasy of man and of science:

> In other words, scientific knowledge results from the passage that changes a cause into a thing and a thing into a cause, that makes a fact become a law, de facto become de jure . . . Consequently, the relation of law to fact, of contract to world, which we noted in dialogue, rivalry, and conflicts, renews itself unchanged in scientific knowledge: by definition and in its real functioning, science is an ongoing relation between the contract uniting scientists and the world of things. And this relation between convention and fact, unique in human history . . . has not been given a judicial name. It is as if the verdicts of humans coincide with those of objects. This never happens except in miracles and sciences. (Serres 2002, 22)

Feminists and other minority political activists have long been dubious anytime the word 'natural' is raised within the social contract. 'Natural' used in a scientific context is always, as Serres states, being used in a judicial context as well. It is frequently the last weapon in operations of classification, biological and ecological essentialization, and control over all living things. For other political activists, particularly ecosophists and abolitionists it emphasizes a further disdainful practice in the Order of Mars – speaking for/of/about the absolute unknowable other in legislative terms between scientists and legislators and thus denying the other a voice because factual natural utterances – the screams of dying animals, bearing witness to devastated environments – do not register in the scientific judicial lexicon in spite of their horrific and clear expression of the destructive effects of the human. The invocation of the word 'nature' within the social contract is utterly oxymoronic as it presents us with an ever-tightening expression of the social contract where accountability and personal investment in the claims is further extricated while those humans who benefit from said claims grow ever more powerful. We are left with two observations concerning death. First, the human drive for knowledge, which Serres sees as homogenous in its proclivities in spite of claims to epistemic diversity, where sciences, religions, economics and judiciary operate under the same conditions towards the same goals, is a war on the equilibrium of individual lives of all species and relations between all lives. Second, the world accessed through nature as a positive chaos deploys constant death in the reconfiguration and turbulent nature of entities as they exist within relations with all other entities and for which there are no laws or absolutes of object or relation, before or after. Clearly the first death is one which is antagonistic to life, while the second is one which is antagonistic only to atrophy and is therefore necessary for the vanquishing of the social contract. In this sense Serres already posits the human as harbinger of death and the death of the individual as a classified atrophied subject as against living relations. If this is so then perhaps the two operations I will posit – abolition and extinction – would be effects of thinking life through the Order of Venus rather than aims or goals?

> Man milks the cow, makes the steer work, makes a roof from the tree: they have all decided who the parasite is . . . the farmer of the cow, the carpenter

> of the roof and the priest who kills the steer are not great people. History says so without symbols, without translations or displacements. But history hides the fact that man is the universal parasite, that everything and everyone around him is a hospitable space. Plants and animals are always his host; man is always necessarily their guest. Always taking, never giving. He bends the logic of exchange and of giving in his favour when he is dealing with nature as a whole. When he is dealing with his kind he continues to do so; he wants to be the parasite of man as well. And his kind want to be so too. Hence rivalry. Hence the sudden, explosive perception of animal humanity, hence the world of animals of the fables. If my kind were cattle, calves, pigs and poultry, I could quietly maintain with them the same relations I have with nature. (Serres 2007, 24–5)

Anthropocentric perception of the world is inevitable when the concept of the human as an atrophied mode of subjectivity is maintained. Non-humans, organic and inorganic, animal and vegetal are subsumed by anthropocentric logic into functioning metaphors, whether within life sciences or deliberate fictions. Anthropocentric knowledge of the world is a human expression of the world *for us* rather than a curiosity about being within a world as part of its multiple relations. Anthropocentric knowledge of non-humans converts the object as a life into an object which is indexed based on its use value for the human. When a non-human is studied in relation with other non-humans human modes of understanding preconceived relations persist in observations and are evaluated as such. Anthropocentrism allows for nothing outside of the righteousness of human perception, whether masquerading as logic or morality. Human parasitism does not explore the world or expand the human world. It constricts the stunning diversity of all organisms and ecosystems into a metamorphic, flexible, adaptable but ultimately restrictive anthropocentric mode of apprehension. Nothing falls away. Everything can be included in this restricted vision however aberrant, and what is aberrant while being named natural is suitably punished for their aberrance through the determination of their use value and treatment by humans, both as material entity and as metaphoric symbol of anthropocentric qualities. This is from where sexism, racism but also speciesism comes. When man is the universal parasite there is no species in the sense of difference. Species is an index of symbolic and use value, culturally and temporally contingent. Species refers purely to the human's capacity for each specific parasitic taking from a species' members possible. Neither animal rights nor welfarist politics has concern for the non-human other. Both exhibit concern for human perceptions of the spectacle of suffering, symbolic value (from Oedipal pets to endangered exotics) and an ablation of the shame of knowing what we do to non-humans is neither necessary nor ethical. Industrialized systems such as slaughterhouses, dairies, zoos and all parasitic use of non-humans are nothing but war machines against nature. The rhetoric of 'humane' treatment is a synonym for parasitic activity vindicated through anthropocentric social contracts between humans agreeing or disagreeing with other humans as to how far their guilt can be excused. Nature is excluded. The non-human other is excluded. Discourse on the treatment of non-humans is a social contract between complicit parasites without the consent of the others involved. There is no natural contract with the silenced other and the hyperbolic contract between humans becomes a judiciary alleviation of accountability more than an evaluation of real-life experiences of real lives. Similarly the contemporary trend in posthuman theory to utilize so-called animal relations as novel ways of understanding emerging human assemblages driven by media, technology and biosciences parasites from the otherness of non-humans without gifting anything in return, similarly evacuating the absolute unknowable alterity of non-human life from its self and utilizing it for imagining human futures. It is interesting that this turn to anthropocentrically perceived non-human systems comes at a point where Serres (2015, 9) states we are in crisis due to the hyper-individualism of millennial life, as we have lost our many and varied 'belongings to'. The crisis of the loss of human belongings sees us turn to nature to reconfigure how to belong again, but problematically via abstracted (both from the non-human source and from material corporeal collectivity) systems. Serres emphasizes the loss of the body is integral to this crisis (11, 69) and attending on corporeality is a means by which a bridge can be built from the individual to the collective, which bypasses the abstraction of the individual as a quantity and where the body indicates the qualitative nature of the self. Bridging through corporeality could also be extended to non-human life with which humans are in proximity. Through an understanding of non-human life as being something which may belong to another genus but which is nonetheless a unique quality of organic emergence, speciesism is prevented from abstracting that life into its human parasite use value. Echoing Spinoza's definition of ethics as opening the world for the other to expand expression and be affected in a way which further expands freedom, the bridging of abstraction to corporeality is a reminder that fact is flesh and no amount of anthropocentric argument can alter the effects of constriction or violence the parasitic use of another's flesh causes. A sense of belonging with nature, without knowing or regulating what that belonging means in terms of hierarchy, control and parasitism, may be a bridge to opening to the natural contract. This belonging requires something different to knowing differently. It requires the refusal of the compulsion to know. It demands Serres' greatest gift to philosophy – grace:

> Whoever is nothing, whoever has nothing, passes and steps aside. From a bit of force, from any force, from anything, from any decision, from any determination . . . Grace is nothing, it is nothing but stepping aside. Not to touch the ground with one's force, not to leave any trace of one's weight, to leave no mark, to leave nothing, to yield, to step aside . . . to dance is only to make room, to think is only to step aside and make room, give up one's place. (47)

In his discussion of grace Serres repeatedly points to grace as belonging to the man who is nothing, blank, a dancer who is naked, who owns nothing and is nothing. This man is jubilant and laughing precisely because he is extricated from the world of signs and symbols, of objects defined by their relation of capital with him. All he is is a body and his abstraction is through the body – abstract because his is a body in movement without words or language, with no signifying attachments. This is a body that is natural in that it is neither signified in itself (hence my preference for 'it' over 'him' although Serres oscillates the pronouns 'he' and 'she' in this section) nor stratified within a social system. Modes of communication must come through unique and specific relation because there is no language to share beyond the shared expression of flesh it may encounter with another fleshly body, of any species or type. Most importantly, this is a body echoing with a violent history of parasitism and destruction, so its coming and going should produce affects which open the world rather than continue the patterns of Mars human occupation perpetrates upon nature. This is a material body which belongs within a Spinozan ethical relation because this body is 'free, perverse, insane and rational, capable of anything. And man is nothing. He is naked. I believe that one can still think of man, universally. But this universe is empty and blank' (Serres 1995, 47). This is an immanent body because it is not limited by its history or acting in order to produce a particular future. It is a vitalistic body because it is free and its grace produces freedom for and in other bodies. However while Serres states the gracious body is 'not even an abandoned memory' (47) the need for becoming gracious, the need to step aside, does come from contemporary devastations of the natural world.

Grace involves a stepping aside which can be clearly seen in the opening of spaces for the other – non-human, ecosystematic – through a forsaking or refusal of the use (always exploitation) and parasitivism of the other. It seems peculiar that activism occurs simultaneously with this absenting; however this is not a paradox for Serres. While the corporeal affect of exploitation of the other or our parasiting of the Earth is a clear material indication of the need for an equally material stepping aside to open out untold spaces of expressivity for others, our relationship with knowledge which produces our capacity to apprehend and express materiality, which thereby converts to our vindication for our actions towards material alterity, is even more urgently in need of grace because it has come to replace materiality itself. From his discussion of grace Serres (1995, 47) states, 'I believe that man is blank and undifferentiated. Man has no instinct, man is not determined, man is free, man is possible . . . I believe one can still think of man universally, but this universal is empty and blank. A universal perverse, man is without attribute.' Gracious stepping aside involves our divestment of the hyper-signification of our own selves, which spills over into a hyper-signified universalization of the Earth *according to us*. While it is clear that the dominant defines the other in order to vindicate parasitism and subsume the freedom of the other to express themselves and the world, defining ourselves sets a template by which we can only know the world and by which we know our knowledge of the world to be right because it is the very knowledge that brings ourselves into being (the conversion according to Serres from judicious opinion to scientific absolute truth). Grace in activism involves the more obvious stepping aside through the refusal to exploit the bodies of others but also the refusal to think the other which for Serres explicitly involves the refusal to think the thinking of the self. This task is what he sees as the primary task of the philosopher and of what I would see as the gracious philosopher. Serres is invested in the possible as the drive and the affect of the philosopher's opening of thought towards freedom or grace. For Serres (1995, 22) the possible is not potential (what we pre-perceive as available or likely) and 'it is the very reverse of power'. He states:

> It is the very function of the philosopher, the care and passion of the philosopher to protect to the utmost the possible . . . the philosopher is no longer right or rational, he protects neither essence nor truth. It is the function of the politician to be right and rational, it is the function of the

> scientist to be right and rational . . . [the philosopher] wants to let the possibles roam free. Hope is in these margins, and freedom. (23)

It is important to remember that according to Serres the scientist and the politician are 'functionaries' along with many other majoritarian human subjects, of truth, logic, reason and rationality. It is critical to remember however that the content of their expression – their claims, speeches and all significations they impose – have no relation with truth and its associated values. Their functionary status overrides and attests to their enunciations so that the functionary replaces the content and no matter what absurdity or limits to freedom are uttered from these positions, the content will be followed because the position itself as truthful has replaced an investigation into its content. Arguments and disagreement of course can ensue but are limited to disputes based between the same functionary status. The other, that who is neither us nor comparable to us, to which we must be gracious, is excluded from this system. This is because the other is the possibility of unknowable otherness as in the evidence of infinite possibility itself – of other positions, of other paradigms, of silence, nothingness. The unknowability of the other is voluminously affective but without a relationship with truth or falsity, right or wrong. The existence of the other is sufficient enough. The persistence of the future in generations humans produce is a similarly functionary position. The seemingly innocuous, innocent child with whom 'hope' is invested is a human fantasy that relies on the faith already invested in scientists and politicians to offer us a better truth. But the child is a symbolic icon of the persistence of human perceptions as truth and the devastating material consequences of the perpetuation of the human species on the Earth is not included in this version of hope which is abstracted from material affect and belongs more to the faith which resides with the functionaries, especially the priest (for logic the priest politician, for truth the priest scientist). The philosopher should try to navigate the social contract but ultimately steps away from its constrictions of singular expression. The philosopher is the one speaker unafraid to step away from the over-valuation of self which constitutes human existence. The philosopher attempts grace through ethics. From a Spinozan perspective the most simple system of ethics opens up the freedom of the other to express through expressing with as many non-imposing possibles as are available and acknowledging these all have various affects. Certainly enunciations of truth do not allow for the expressivity of the other but rather place them within the singular human categories from which they can neither express nor be free. The opening of freedom for the other is grace as an actual stepping aside through abolitionist practices of activism and anti-natalism, and the philosophical exploration of an ethics towards the natural contract which steps aside from regimes of signification of self and other, indeed of the world. Serres (2007, 213) advocates the cessation of application, the opposite of philosophical grace – to apply knowledge onto the other 'analyzes, paralyzes, catalyzes', transforming human social contracts of truth without engaging with the natural contract. The violence of knowledge (or the Order of Mars) is combatted with the Grace of Philosophy (or the Order of Venus). While we cannot step aside from our residual humanity and simply enter the natural contract (which would be fetishism and likely destructive considering the history of the human) we can create a voluminous gap which teems with possibles and we can exercise our grace in welcoming the silences, the illogics, the cacophony, the imperceptible multiplicities as navigational tools to open up the world to itself, by its own means and in its own way while withdrawing our violence towards and presence from it. This is where grace changes knowledge to love – perhaps a very beautiful gift of the definition of the philosopher Serres gifts to us: 'Love has just been defined as an intermediary . . . it occupies the middle spot between knowledge and ignorance. Love can be thought of as being among the fuzzy subsets. He is the included third. He is between' (246). The gracious human who refuses through activisms such as abolition and anti-natalism, as well as the philosopher caught between the natural and social contract, can express love through the ways they are caught up as the in between – traitors to their species yet within a world where their affects are still felt. The philosopher is, after Serres' definition, treacherous to the transcendental over-valuation of the essence or meaning of the subject (perhaps in this way treacherous also to the history of philosophy and certainly a pariah in contemporary discourse) and is set on ending reproduction by ending their role as enunciator who perpetuates future generations of truth and logics and fetishizing, exploiting or ignoring the other in the search for the transcendental Man. Step aside then, from using the other and from reproducing the same – the job of the philosopher and the activist share much.

Grace deposes the social contract but from an activist perspective there is a care element required when we become gracious. Stepping aside for nature resolves the activist argument for abolition of use of non-humans but human intervention has wreaked havoc in ways which involve care in order to create a more Venusian state of equilibrium in organisms and the world. Strategies for this care must also be gracious in that they do not persist along modes of operation of relations between humans and non-humans that have gone before, they do indeed require us to abandon our memory of how humans relate with each other and with non-humans, to formulate specific bridgings between corporealities and ethical acts towards opening the world to biodiversity. Serres closes the fissure in abolitionist activism its critics have received between abolition being about leaving non-humans be while simultaneously helping those non-humans who suffer by showing that the dance of activism is a leaving be yet acknowledging encounters occur and relations should develop within these encounters based on the specifics of each relation and the organisms involved. Critics of abolition activism have always fought via the Order of Mars anyway, as they retain the argument within the social contract between humans. Becoming a gracious activist may involve leaving one's species, becoming dead as human in order to create a belonging with another without definition (what Spinoza calls common notions – commonality without equivalence or awareness, only imperceptible resonances in shared being). But human grace can go further still. We can cease to be altogether.

A vitalistic ethics of extinction embraces cessation of taxonomy and species, while simultaneously requiring an embracement of the dynamic material reality of all life and lives as they are now. Considering both the mass and diversity of life on the planet this is already a formidable form of activism, both conceptually and actually: conceptually in that each engaging encounter with another life requires imagination and attention to the specifics of the relation independent of preordained structures of object and subject or self and other, and actually because by forsaking those preordained structures the effects of each relation are not deferrable to previous or future records of encounter. We can predict them only insofar as we can hope the other's capacity for expression will be opened by our own openness to alterity itself. The mingling of grace as a leaving be and a duty of care requires a certain kind of what Serres might call 'clinamen activism' where turbulence is mingled with equilibrium to afford change in the terrain of potentiality for the other (Baruch de Spinoza's *potentia*) without imposing new legislations in relations (*potestas*). Considering the expansive size of such a project any argument that the human species *needs* to continue (taking into account with the contemporary generations it would likely do so for at least another century) seems overwhelming. The human desire to breed, for whatever reason from the secular move to immortality through progeny to the biological essentialism of the reproductive 'clock', is simply incommensurable with a desire to ethically create a natural contract. Breeding as a concept has been radically converted in its industrialization to a driving force of human hubris in exertion and control from factory farming of non-human animals to the human reproductive biosciences, all of which have resulted in the industrialization of life itself as commerce, as abstracting dividuated vitalism from life as currency, be it the currency of the cow murdered for consumption to the currency of the IVF biological child as authentic copy of its parents. Human extinction involves the care of human lives here and now, of which there are already too many deposed and disenfranchised, and the cessation of the manipulation of the reproduction of non-human animal lives designed purely for torture or murder for human use. When reproduction is removed, all interaction becomes production – creative adaptable imagination towards opening space in an ethical relation while acknowledging the finitude of the Earth's resources. Abolitionists and extinctionists have heard all of the common arguments against both (commonly referred to as 'vegan bingo' or 'antinatalist bingo' due to their trite repetitive predictability), however almost no serious studies which advocate either have been offered due to the massive investment capitalism has in both the abuse of non-human animals and the validation of human reproduction as necessary. Interest in either – from individuals or companies – inevitably ends up being self-interest (or vindication) and so even a scientific study of the viability of either becomes, as Serres states, a judiciary agreement between humans with shared ideological agreements. The refusal to give away human privilege – which in the case of human reproduction is only the *idea* of a human yet to come – seems to be a block in the journey towards a celebration of life as the interaction of all living entities and their relations through a natural contract:

> Back to nature then! That means we must add to the exclusively social contract a natural contract of symbiosis and reciprocity in which our relationship to things would set aside mastery and possession in favour of admiring attention, reciprocity, contemplation, and respect; where knowledge would no longer imply property, nor action mastery, nor would property and mastery imply their excremental results and origins. An armistice contract in the objective war, a contract of symbiosis, for a symbiont recognises the host's rights, whereas a parasite – which is what we are now – condemns to death the one he pillages and inhabits, not realising that in the long run he is condemning himself to death too. (Serres 2002, 38)

The vastness of this project is a remaking and a rethinking of the world – where thought is without established 'knowledge' – an ahuman understanding of living and life within the world as it is in its already immeasurable relational combinations, locally and globally, infinitesimally and sometimes seeming infinite. But immanence and finitude are two key features of why the anthropocene as it is known is being called the death knell of the Earth, or the last epoch, due to human life's persistent abstraction of action and expression from effect, thus causing changes in the world which are unsustainable. The desire for 'sustainability' maintains the human at the apex of what we see as needing to be sustained, primarily our globally dissymmetrical social contract at the expense of nature. It retains the compulsion to squeeze the last bit of life from nature for the benefit of humans. A gracious form of parasitism not only recognizes the host's rights, but also that we cannot recognize what the host wants or needs, because it does not demand the host speak in the language of the parasite in order to achieve equality. 'This is where parasitism reappears', writes Serres (2002, 37), 'The Declaration of the Rights of Man had the merit of saying "every man" and the weakness of thinking "only men", or "men alone". We have not yet set up a scale in which the world is taken into account in the final balance sheet.' While Serres advocates all objects (all lives the human decries as objects for not being human) become legal subjects, humans cannot be the keepers of legal jurisprudence, as this would relegate nature to proving its equivalence within the symbolic systems of the human speaker. This has of course always been an issue for feminists of difference (rather than those wishing for equality) and for all victims who cannot speak the language of the master, what Lyotard calls the *differend*. The human silenced and stepping aside creates a space of contemplation and respect for the other without demand or imposition, utilization or assimilation. This prevents the many illogical logics the human sciences perpetrate on the natural world to vindicate their violation of it:

> Method in the human sciences, which deal only with relationships, apes police and inquisitorial suspicion. It spies, shadows, sounds hearts and minds. It asks questions and remains suspicious of the answers, it never

> asks itself whether it has the right to act as it does . . . The human and social sciences describe theories even more underhanded than fraud, more duplicitous than cheating, in order to outsmart their object. Here everything becomes possible; a cow is a woman or a god a bull, even the identity principle is unstable. Reason watches while reason sleeps, reason sleeps while it watches, a hellish world of relationships in which stability itself fluctuates. (Serres 2008, 42–3)

Humans have relationships with things because they are parasites – what are you to me, what do you mean to me, what can I do with you? When challenged as to the necessity of abuse of non-human lives, be it through their incarceration for entertainment, murder for consumption or torture for 'research', the human will make the challenger account. Similarly the breeder, following the default heteronormative linearity of birth-marriage-breed-die, has a relationship with their own perceived immortality in breeding and their own personal human narrative, the effects of themselves and their progeny on the future of nature is not seen as a problem, their habit in following human normative narratives is converted to a right. The rights of humans, even in the face of the crisis of the anthropocene, override the fact of overpopulation and the murder of non-human animals. The human as an idea and an agent brings only death. Forsaking relationships for relations, forsaking human subjectivity towards becoming ahuman, is the antithesis of death even though it involves the cessation of the human as a species and concept. It produces new modes of interaction, it opens both the ahuman and the world to encounters with the natural contract, a thoroughly demanding and joyous mobility which rethinks itself. Ultimately it will see the former parasite care for its hosts until the parasite ceases to be. The demand for the imagination it will take to manage human extinction while managing the damage done to the hosts is a gift we can give ourselves in creating new ways of being which make us all philosophers of immanence with our immediate surrounds. Metaphysical questions which inevitably end in nihilism due to being unanswerable such as 'who am I?' or 'why am I here?' give way to the more Spinozan question of 'what can I *do*, here and now?' Each body's capacity to act is unique and therefore each body is necessary. Our doing involves as much not doing – not murdering, not breeding, even not being alive – as it does action, so capacity is defined by uniqueness of ability, not ability measured via a Vitruvian template of human majoritarianism. Serres gifts in his critique of the social contract and his desire for a natural contract the emergence of the world as a material fact, a gift that human signifying systems have obscured, a gift that denies human lives their enjoyment of immanence through constant deferral to pasts, futures and abstract capital objects, a gift that could allow non-human lives to flourish by simply being allowed to be without subjugation. In these ways Serres is rigorously useful and hopeful for two activisms which still seem radical or extreme – abolitionists and extinctionists – but which this epoch, whether official or apocryphal as the anthropocene, needs urgently to manage bearing witness to the devastation the human has wreaked.

> The tree and cow told us that man never returned or recognised the gifts of flora and fauna. He uses and abuses them but does not exchange with them. He gives food to the animals you say. Yes sir he gives the flora to the fauna, fauna to the fauna . . . what does he give of himself? Does he give himself to be eaten? The one who does so will utter a timeless word. One word, *host*. (Serres 2007, 82; emphasis in the original)

References

Serres, Michel. (1995). *Genesis*. Trans. Genevieve James and James Nielson. Ann Arbor: University of Michigan Press.

Serres, Michel. (2002). *The Natural Contract*. Trans. Elizabeth MacArthur and William Paulson. Ann Arbor: The University of Michigan Press.

Serres, Michel. (2007). *The Parasite*. Trans. Lawrence Schehr. Minneapolis: University of Minnesota Press.

Serres, Michel. (2008). *The Five Senses: A Philosophy of Mingled Bodies*. Trans. Margaret Sankey and Peter Crowley. London: Athlone.

Serres, Michel. (2014). *Times of Crisis*. Trans. Anne-Marie Feenberg-Dibon. London: Bloomsbury.

Serres, Michel. (2015). *Thumbelina*. Trans. Daniel W. Smith. London: Rowman and Littlefield.

Serres, Michel. (2018). *The Birth of Physics*. Trans. David Webb and William Ross. London and New York. Rowman and Littlefield.

10

A new culture to suit the world

Interview with Michel Serres by Janina Pigaht and Rick Dolphijn, translated by Joeri Visser

And we went around saying: of what use are the humanities, cultures, or religions? Answer: they're useful for not dying; they've been useful up to now for keeping us alive.

MICHEL SERRES, STATUES, 7–8[1]

The following interview took place in the apartment of Professor Serres in Vincennes, on 29 August 2014. After letting us know that, sadly, he couldn't deliver his keynote (because of medical issues) at our conference in Utrecht, the Netherlands, Professor Serres kindly invited us to his home, so we could film the talk there and play it at our conference. A blue gate, in between typical neoclassical Parisian residential blocks, with only a tiny handwritten sign 'Serres' above the doorbell, led us to his nineteenth-century home. In the heart of Vincennes, once a suburb of Paris but now completely woven into the city, it was an oasis in every sense of the word. The walls overgrown with books on its inside, the luscious green grass fields, flowers and trees outside, it was the best of all possible worlds. In absolute silence, with the yellow sunlight coming in from the low window on the left, emphasizing the art history book collection that enlivened the wall behind him, the scene turned into a Vermeer painting as well. And this was all before the interview had even started.

When the conversation commenced, Professor Serres generously shared with us his life story. A life in philosophy that is a philosophy of life as well as a life of philosophy. It is a life, as we will soon find out, that gives rise to a wholly other world, a wholly other university and a wholly other society. It is a life that demands us to rethink the relation between the sciences and the humanities, between the currents of our day and age and the undercurrents that we so often refuse to see.

It preceded this book, and it serves as the perfect conclusion to it.

Q: Professor Serres, perhaps you could start this interview by introducing us to your life in the academy.

Serres: At first I was a scientist; that means that I studied mathematics and physics. And, at some point, I took a new direction in choosing the Humanities, and I chose the Humanities because of a particular event or a particular encounter. I was born in 1930, so I was 9 years old at the outbreak of the Second World War and 15 years old at the end of that war. The most important event in my life as a scientist was the dropping of the atomic bomb on Hiroshima and Nagasaki. That explosion was decisive for a philosophical reason: before the Manhattan Project,[2] we were all convinced that science was only good: it was called scientism. And suddenly, there was an extraordinary crisis of conscience; science turned out to be bad, because it caused the death of thousands of men. And so, the scientific community was gradually seized by that problem. I know many scientists, physicists, for example, who became biologists because they wanted to move away from atomic science.

I was 15 years old at that time and began doing science, and I had been thinking a lot about the implications of Hiroshima and Nagasaki. That brought me to the Humanities. Since the question arose within science, it could only be resolved by philosophy. And therefore, that particular event made me shift from science to philosophy. Thus, I did not really choose the Humanities; they were imposed on me, because this event was decisive for my entire generation, the generation before me and the generation after me. Later, of course, questioning science became common; within chemistry, biology etc., but it was always the same ethical questioning of science, of all the problems affecting the notion of progress. And so it was necessary to consider this question in the light of philosophy and I therefore engaged in the arts. And my whole life, the Humanities have been a kind of bridge for me between the hard and the soft sciences.

Q: Can you explain the difference that you make between the hard sciences and the soft sciences?

Serres: Hiroshima and Nagasaki raised questions within physics, within thermodynamics and mathematics. So, the question was raised within the hard sciences: physics and mathematics. But the same ethical or moral questions were then raised within chemistry, because there had been accidents in Seveso[3] and other places. Then biology was questioned about its genetic engineering. And so, the whole field of sciences became involved in that question. And we could only resolve it with the aid of philosophy. And therefore, all my life, I've tried to hold science with one hand and arts with the other. For me, Humanities is this kind of bridge or connection; a relationship between science and arts. Besides, scholars from the Renaissance were scientists as well as philosophers.

Q: How would you position yourself as a scholar? Do you consider yourself a Humanities scholar?

Serres: When you talk about a position, one must define a special place that I inhabit within the totality of knowledge. Well, I have no position. Since I'm on the bridge between science and arts, I have no position. I am precisely in the relationship between the two. And I believe that the great philosophical tradition is about having no position. Let me give you an example; when you go to the doctor, there are general practitioners and specialists. Your doctor is a generalist who says: 'Go see the cardiologist, the neurophysiologist or the neurologist.' These are specialties; the doctor is a generalist. By analogy, philosophy is the generalist and the other disciplines are the specialties. I am a generalist.

Q: The generalist perspective, which marks philosophy (and the theoretical Humanities), is perhaps at odds with the current emphasis on methodologies that one finds also in these parts of academia.

Serres: Suppose you are cooking. In your cookbooks are recipes for cakes, roasts etc. These recipes point out a method: 'take the eggs, the salt, the flour and make a dish'. The method assumes that this dish was already made earlier and is thus known. So, having a method is also already knowing the outcome. Let me be clear: there is no method.

Do you know the word 'serendipity'? The word 'serendipity' is taken from an English novel.[4] In France, we have an equivalent. Let me tell you a short anecdote from Zola's novel *Au Bonheur des Dames* (*The Ladies' Paradise*). In this novel, Zola tells the story of the man who invented department stores. When the man invented department stores, he classified the products that he sold in a methodical order; there were vegetables, fruits, shoes etc. These products were all arranged and a year later he realized that the revenue was peaking. That is to say, he constantly had the same revenue – there was no growth. Then, one day, he had an extraordinary idea; he disrupted the order of the products. That is to say, he mixed vegetables with perfumes, shoes etc. So, the buyer who was returning to the store and who wanted to buy shoes suddenly also saw vegetables, perfumes. So instead of one product, he bought two or three products. And consequently, the revenue was growing again. Therefore, if you have a method, you will manage to make your cake, but you manage to make something that has already been found. One can only discover through serendipity. A method is good to learn, to form oneself or to teach. But not to discover, to find or to invent. No method, but serendipity.

Q: What does this mean for doing philosophy today, for thinking the contemporary and what may follow?

Serres: I have a goal in my life and I believe that the Humanities should also have this goal. We are currently experiencing a considerable revolution. The world is changing a lot. For example, the balance between cities and rural areas has changed and life expectancy has changed. Did you know that a 60-year-old woman is further away from death than a newborn was in the seventeenth century? Thus, body, life, death and humanity have all changed. When I was born, the world's population was less than two billion people. In my life, this is multiplied by two twice; there are nowadays nearly eight billion people. So, it's not the same world, not the same life and not the same body. But we still have the same institutions, the same politics, the same governments and the same rights. That's an untenable situation. On the one hand, there is the world – as it is; the human world – and on the other hand, the historical processes and changes. There is a considerable change that is comparable to what happened when the word 'Humanities' was coined. Humanism was developed during the Renaissance; when there was a shift from the Middle Ages to the Renaissance. Exactly as in our times, everything changed; it was necessary to invent a new culture, new institutions, a new way of being together etc. And therefore, the goal today is trying to prepare this new culture; these new humanities. This is considerable and new but exciting and interesting work.

Q: Does this new world demand a new university?

Serres: I believe that the university has to change, because our academic model dates back to the nineteenth century; it's practically the German heritage of the early nineteenth century. It also dates back to the universities of the Middle Ages and it may also date back to the Greek institutions; the Hellenic paideia. That's all very fine because there is nothing better for training researchers. But today the university is shaken by the earthquake that is the contemporary and modern world. And there is again this discrepancy between the contemporary world and the university which was made in a world that is not the one it has become today. So, the university must certainly change. But how? I do not know yet. But I think the question should be how to change the university today. Let me give you an example from *Petit Poucette* (*Thumbelina*) – a book I wrote on digital issues. Architects came to see me asking: 'We are designing a university. And, since you wrote this book, tell us how the university of the future should be designed.' It's impossible to answer them and there you see the problem. So, what affects the university also affects governments, politics, democracy etc. It's a tension between the world as it is and all the institutions that are relativity obsolete.

Q: So the world in which we live, and the world in which we will live in the near future, demands a generalist university.

Serres: In Paris, the government has decided to set up a new university in the north of Paris that will be devoted to social sciences and a university in the south of Paris devoted to hard sciences. In the south of Paris, there will be mathematics, physics etc. and in the north of Paris, there will be sociology, psychology etc. This is complete nonsense, because, the future of knowledge – the Humanities – is about mixing the two. Nowadays, a sociologist, for example, who does not know mathematics makes no sense. So, I believe that the big problems can only be resolved in a joint alliance between social sciences and hard sciences.

Q: Within the social sciences and especially within the Humanities, this 'opening up' to the other Faculties, is already taking place (with some successful and some unsuccessful outcomes). But is this 'opening up' taking place within the sciences?

Serres: Science education suffers from a deficit because scientists do not know the history of science and they hardly know the implication of science in the contemporary world. It is therefore very important for a physicist to know a little about sociology in order to know what he is doing. I once prepared a text for scientists. In ancient times, when medicine became scientific, Hippocrates wrote the Hippocratic Oath. Nowadays, doctors still take the Hippocratic Oath. I invented an Oath for Scientists[5] in order to try to make scientists aware of the fact that they operate within a world that is enlightened by arts and that they cannot go into this world without the Humanities. I tried to formulate this, because it's necessary.

Q: Are these alliances about to happen then?

Serres: We started some time ago with the so-called MOOC,[6] that means online education. It's already old, because teaching at a distance exists already in almost all countries for a long time. In Australia, Canada, Spain and even France, there has been education at a distance for people who couldn't come to university. Thanks to the new technologies, online teaching can take on new dimensions. There are obviously many universities that began with failures, but also many that started with successes. It's not yet well developed, but this will probably be the direction in which we are moving.

Q: To conclude, if you could choose a name for the faculty in which you were working, what would that be?

Serres: It should bear the name of someone who understood the contemporary world and who tried to invent a new culture to suit that world. Maybe Erasmus or Montaigne, because the right model is found at the time of the Renaissance. But I want a name that echoes the alliance between science and arts. Perhaps Cervantes. Why? Cervantes is very interesting, because there are two characters: the intellectual – who sees the world only in the books and therefore attacks windmills – and the other character who is down to earth: Sancho Panza. There is the intellectual and the people and it's about the alliance between the two. That would be the name: Cervantes.

A part of this interview has been included in The Humanities in Europe – or European Humanities – video interview series, a project by Prof. Rosi Braidotti, director at the Centre for the Humanities at Utrecht University (NL), implemented in cooperation with Janina Pigaht, Centre for the Humanities Artist-in-Residence. It can be seen online at: https://www.youtube.com/watch?v=SnE1AHjIONQ.

NOTES

2 The virtue of sensibility

1 Serres notes that this structure is found in cybernetics (Serres 1995, 42). See also *Times of Crisis* (Serres 2014, 28).

2 Serres describes this in his reading of Lucretius (Serres 2018, 147–48). The idea of physical laws as emergent regularities is examined in depth by Roberto Maneira Unger and Lee Smolin (2015).

3 As such, the language left over is similar to the written word cut off from the voice. Serres promotes the spoken word, and the written word that in its music still bears the traces of its animation by the voice.

4 'Sometimes, from the chaotic tohubohu and fluctuations, at the edge of my "consciousness", from this background noise and its tiny signals, imperceptibly fine, the fragment of a timid melody may arise, unheard and come to my hearing: ululation, lamentation, hymn, cantilena, ballad, melopoeia . . . It drags me from my tears or uplifts me with tenderness. I desperately strain my ears, inner and outer, to understand its birth' (Serres 2009, 48).

5 In the end, Orpheus twice escapes death only then to meet it twice. Having been torn apart by the Thracian women, he passes to the underworld and to death in language.

6 Moreover, the harmony he creates does not reduce the multiple to one. Rather, it is 'a way of encompassing everything' (Serres 2008, 138): all possible senses are held there (123). This recalls the way that as early as 1966 Serres (1968a, 94) described the emergence of form as dependent on the fusion of all possible times.

3 Cosmoliteracy and anthropography

1 'We have lost the world. We've transformed things into fetishes or commodities, the stakes of our stratagems; and our a-cosmic philosophies, for almost half a century now, have been holding forth only on language or politics, writing or logic' (Serres 1995, 29).

2 Cf. for an introduction to this motif in Kantian philosophy, Grier (2012).

3 In a short preface to his book *Rome, First Book of Foundations* (2015), Serres describes his book as a first approach to history in the objective, scientific terms he deems adequate to address this novel form of global, objective violence:

> The shaking that grips me upon starting a book on history isn't from fear; I'm not afraid. And yet, here, terror reigns, murder, blood and tears, constant iniquity. I know that we never encounter any social system that's just; I've rarely known, living or dead, any powerful man who was good. The shaking gripping me is not from fear; it is, if I may, from logic. It would be an exercise infutility if a philosophy formed from its instauration by the rigorous and precise concepts of the sciences of the object brought its practices into the unstable cloud of time. It either wouldn't understand, or it would be formed with shaky outlines. History is fuzzy and vague, but it was precisely the sciences of the object that prepared me to think this shakiness with exactitude. So here I am on the terrain of terror, for the first time, finally ready, despite my anxiety. This century we have new tools. Here they are.

4 That is in terms of cryptography, via the position of the third, the interceptor, the parasite. In this, Serres' approach to communication, and the physicality that manifests in communication technology, diverges categorically from any kind of a multiple-nature approaches that seek to restore a balance between them. Indeed, wherever system theory, logical metalanguage approaches, higher order cybernetics approaches speak of 'balance', we have to think 'contract' if we want to understand Serres' approach.

5 Such a notion of 'virtual actuality' is different from Gilles Deleuze's approach, according to which the virtual is real but not actual. Deleuze wants to decouple the virtual from any positive notion of possibility, and so does Serres. But Serres' approach is one that considers a substantial notion of chance, that must be thought of as an invariance that underlies the countable possibilities of what in probabilistics is called a 'state variable', or 'random variable'. Such a substantiality of chance is closer to a quantum physical substantiality than to one compatible with a physics of forces. It attributes chance's indeterminateness to a transcendental notion of the objective, not to a subjectivity of particular cognitive agents. The whole point of a 'communicational physics' is to take into account a kind of quantum physical actuality that is at work in his vicarious domain of placeholders. Cf. Crahay (1993).

6 There are interesting proximities between Michel Serres' and Luce Irigaray's work on an ontology of breath and the role of Mary therein (Irigaray 1983, 2010).

7 Cf. the motif of Aphrodite in Serres (2018, 43, 133, 137, 139, 166, 170, 185).

8 For a starting point to pursue an understanding of this, cf., for example, wikipedia.org, on 'crystallization'.

9 What language do the things of the world speak, that we might come to an understanding with them, contractually? But, after all, the old social contract, too, was unspoken and unwritten: no one has ever read the original, or even a copy. To be sure, we don't know the world's language, or rather we know only the various animistic, religious, or mathematical versions of it. When physics was invented, philosophers went around saying that nature was hidden under the code of algebra's numbers and letters: that word code came from law. In fact, the Earth speaks to us in terms of forces, bonds, and interactions, and that's enough to make a contract. Each of the partners in symbiosis thus owes, by rights, life to the other, on pain of death (Serres 1995, 39).

4 Mathematical anamneses

I am thankful to Etienne Balibar, Peter Osborne, Hamid Taïeb, Mauro Senatore, Charlie Clarke and Jaideep Shah for their comments on earlier versions of the following text.

1 Unless otherwise indicated, all translations are mine.

2 Leibniz, Serres argued, devised a *double* method of universality: the first one starts from an *elemental* consistency (the monad, the singular), on which he reads the universal law, while the second starts from the *mathesis* (the monadology, the system) or abstract generality, from which he proposes to derive the particular. In Leibniz's system, these universalities form a *cycle*, meaning that these two methods are fundamentally co-conditional and cannot be dissociated.

3 These texts are: 'Ce que Thalès a vu au pied des pyramides', first published in Serres (1972) and translated into English as 'Mathematics and Philosophy: What Thalès saw . . .' in Serres (1982, 84–97); and 'Origine de la géométrie' 3, 4 and 5 in Serres (1980). Only 'Origine de la géométrie 5' has been translated into English as "The Origin of Geometry' in Serres (1982, 125–33).

4 Serres indicates that this text was written in 1966. It was initially published in 1967 in *Archives internationales d'histoire des sciences* 20, and then republished in 1968 in *Hermès* I, *La Communication* (78–112). Twenty-five years later it was trimmed to figure again as the opening chapter of *Les origines de la géométrie* (1993) under a new heading: 'Differences: Chaos in the History of Sciences' (15–35). I will base most of my analyses on the 1968 version of the text.

5 According to Etienne Balibar, the expression 'history of truth' is at the crux of the debates around logics, epistemology and phenomenology that animated the French philosophical scene between the 1950s and the 1980s. For Balibar, this expression marks the specificity of the French 'moment' of the 1960s, conferring it 'a relative autonomy with respect to its international environment' (Balibar 2004, 23). The

question is addressed at great length in his chapter 'Être dans le vrai?', in Balibar (1994, 163–209).

6 Serres' early 1960s project of recasting the Kantian transcendental took various forms. Alongside his works on mathematics which attempted to historicize the transcendental, he also developed the concept of an 'objective transcendental' (*transcendental objectif*). On this notion, which lies beyond the scope of the present chapter, see Serres (1972) and Crahay's (1988) synthetic commentary.

7 This debate was launched by Cantor, Frege, Russel and Whitehead, and pursued by Brouwer, Dedekind, Hilbert, Atkin, to name a few, from the late nineteenth century to the 1930s. For Serres, this event had not only proven determinant in the history of mathematical theory and subsequent findings, it also marked a point of rupture between 'classical' and 'modern' mathematics. On this topic, see Serres, 'La querelle des anciens et des modernes', in Serres (1968, 74).

8 In a nutshell, Gödel had shown that no consistent theory containing the theory of integers could be complete or entirely proven within that theory.

9 Cavaillès' reflection on the *structuration* of mathematical 'thought' (or 'experience') emerged from the same intellectual space as Bourbaki's 'algebraic structures', in the formalism of the Göttingen school (Hilbert, Artin, Noether) and is thus directly related to Serres' views. See Benis-Sinaceur (1987, 5–30).

10 As a matter of fact, the crisis had been of continuous relevance since the early twentieth century, constituting, as José Ferreiros argues, 'a long and global process, undistinguishable from the rise of modern mathematics and the philosophical and methodological issues it created' (Gowers et al. 2010, 142). In 1960, for Serres, the crisis was still open. Serres interpreted it as a self-reflection of classical mathematics, which had brought about new strata of language. 'La querelle des Anciens et des Modernes', 74.

11 'Is there not', Serres writes, 'a lot of presumptuousness in arrogating the right to talk [*discourir*] about a rigorous language without first settling the language of this discourse?' (1968, 62).

12 Serres, 'Les anamnèses mathématiques', 85–95. Addressing Serres' analysis of the historicity of science in greater depth would require going back to his doctoral thesis. In *The System of Leibniz*, Serres had suggested we interpret Leibniz's conception of historical progress through a variety of mathematical models. According to him, Leibniz's achievement regarding the question of history was that he had not proposed *a* philosophy of history but had instead described a 'schematic dictionary, a formal inventory, drawing a space of choices on which he draws the graph of the history of possible histories' (Serres 2007, 284).

13 This way of reasoning is a clear influence from Bachelard, who writes, in *The Formation of the Scientific Mind*:

It is therefore this striving towards rationality and towards construction that must engage the attention of epistemologists. We can see here what distinguishes the epistemologist's calling from that of the historian of science. Historians of science have to take ideas as facts. Epistemologists have to take facts as ideas and place them within a system of thought. A fact that a whole era has misunderstood remains a fact in historians' eyes. For epistemologists however, it is an obstacle, a counter-thought. (Bachelard 2002, 27)

14 There are two modes of archaeology: intrinsic and extrinsic. Intrinsic archeology is 'the movement mathematic as such, which ceaselessly reactivates its origins and deepens its foundations, by iteration of its internal recurrence, unravelling primitive idealities that were not mathematical and become so by this move.' As such, it is both recurrent and teleological. Extrinsic archeology, however, consists in reading the prehistory of mathematics' *abandoned concepts*, and with these fossils to 'reconstitute the lost genesis of a lost ideality'. Unlike the first, this movement is only regressive: 'the progressive path of effectivity is forbidden and crossed . . . as the ideality it deals with is no longer mathematical' (ibid., 102).

15 As Derrida remarks in his commentary of Husserl's 'Origin of Geometry', the problem of the origin of idealities resides in the ambiguous character of this 'before' and 'after'. In this question, the genetic perspective and the consequential perspective are indeed interlaced in a singular way. See Jacques Derrida, 'Introduction', in Husserl (1999, 55); Derrida (1989, 65).

16 The presence of this thematic both at the beginning (*Hermès I*) and at the end of the series (*Hermès V*) indicates, I believe, its crucial importance. When coming back to it in *The Origins of Geometry* (1993), Serres stressed that he had been thinking and writing about the origins of geometry since 1958.

17 'Ce que Thalès a vu au pied des pyramides', in Serres (1972, 172); 'Mathematics and Philosophy: What Thalès Saw . . . ', 91. The image of the rosetta stone provides Serres with another iteration of this idea (already fully formulated in his doctoral thesis), according to which the real problematic object is not the full determination of the languages (*langues*) at stake, but only the set of relations by which two languages succeed in corresponding to one another: 'Here, no language is unknown or undecipherable, no face of the stone poses problem, what is at stake here is the common edge of two faces, their common border, what is at stake is the stone itself' (189).

18 What is the status of the knowledge implied by a certain technique? A technique is always an application that envelops a theory. The entire question – in this case the question of origin – boils down to an interrogation of the mode or the modality of that enveloping process. If mathematics arose one day from certain techniques it was surely by making explicit this implicit knowledge. (Serres, 'Mathematics and Philosophy: What Thalès Saw . . . ', 89)

19 This idea, which would be later exploited by Latour in his study of the successive translations by which scientific facts are produced, was also at stake in Claude Lévi-Strauss' *Mythologiques* (1964–71). In the latter, he developed a method based on 'systems of transformations', relying on specific myths as 'circulating references'. On the latter notion, see Latour (1999, 24–79) and on Lévi-Strauss' method, see Salmon (2013).

20 'Ma régression ne suit pas le chemin de la tradition, indéfiniment hors circuit, mais le chemin vertical de l'approfondissement mathématique: c'est à partir de là que je réinterprète la tradition' (Serres 1993, 32); this passage was already in 'Les anamnèses mathématiques', 105–6. This 'ruling out' should be understood through his notion of 'external archaeology', cf. *infra*, note 15.

6 The exogenesis of light

1 See his earlier work on Lucretius' atomism: Serres (1977).

2 On the role of the 'third messenger' in Serres' epistemology, see Serres (1982b).

3 Contra Bataille, apparently the cult of the Sun god in ancient Egypt was born after the introduction of the calendar. The abstraction of space (geometry) was introduced to organize and govern the agricultural fields on the Nile delta. The abstraction of time (calendar) was introduced to organize and govern the cycle of agricultural seasons. The cult of the Sun did not emerge then out of metabolic worship but out of temporal abstraction.

4 The expression 'data bank' is used in Romance languages more often than in English. Out of metaphor, it suggests a relation between data and capital that disappears in the term 'database'.

5 For a critique of 'heliocentric slavery', see also Negarestani (2010).

6 This intuition has been extended recently by Francis Bailly and Giuseppe Longo's idea of anti-entropy that differs from Schrödinger's negative entropy. See Bailly and Longo (2009).

7 Originally as 'néguentropie'. See Brillouin (1949; 1950; 1956).

8 See Benjamin Bratton and the idea of the Black Stack in *e-flux journal* 53 (March 2014).

9 Technically solar databases provide measurements of sunlight exposure almost for any area of the world. See the Open Solar Database (www.opensolardb.org).

10 See Wolchover (2014). The idea of dissipative structure was originally suggested by the Nobel Prize winner Ilya Prigogine (1977).

7 Scintillant@the University of Angelic Invention

1 Aby Warburg (1999, 89) said long ago that it is 'possible to trace, step by step, how the artists and their advisers recognized the "antique" as a model that demanded an intensification of outward movement, and how they turned to antique sources whenever accessory forms-those of garments and hair – were to be represented in motion'. He points to Alberti's comments that

> [i]t is pleasing to see some movement in hair, locks, boughs, leafy fronds, and garments. As I said, I myself take pleasure in seeing seven different movements of the hair: hair should twist as if trying to break loose from its ties and rippling in the air like flame, some of it waving in and out like vipers in a nest, some swelling here, some there. Branches should twist upward, then downward, outward and then then inward, contorting like ropes. Folds should do the same: folds should grow like branches from the trunk of a tree. They should follow every movement, rippling, so that no part of the garment is still. (96)

It would be fascinating to track Alberti's series of verbs and metaphors, although we would likely claim that the trunk of a tree emerges from folds, rather than the other way around. 'No part of the garment is still.' My thanks to Florian Knothe, Greg Thomas and Opher Mansour for their suggestions about Botticelli.

2 Audit culture is receiving a great deal of contemporary attention and moves in many historical and theoretical directions. For a preliminary orientation, see Diez (1997), Shore and Wright (1999) and Poovey (1998). I would also want to extend this genealogy to the philosophical discussions of the relationships between quantity, number and truth, which has been part of the scenario since the inception of 'western' philosophy.

3 This is not the place to extend the discussion, but Serres' enactment of a time without a unifying teleology of intention and a clearly marked unitary line between past-present and future, its *percolation*, is yet another style of thinking time that Deleuze – with Aion and Cronos as 'two readings of time' – and Derrida with his discussion of iterability and related terms also enact. When the latter is discussing temporality in Heidegger's *Being and Time*, for instance, he notes that 'here one should not even say *inequality* but *anequality*, inequality presupposing a defect or a shortcoming with respect to a measure of a *telos*, to a common entelechy, to a measure of all things' (Derrida 2016, 208). This, given enough time, would link us back to the data fetish of measurement and metrics.

4 'It may be that education will be constantly improved', Kant wrote in On Education/ Uber Pädagogik, and that each succeeding generation will advance one step towards the perfecting of mankind; for with education is involved the great secret of the perfection of human nature. It is only now that something may be done in this direction, since for the first time people have begun to judge rightly, and understand clearly, what actually belongs to a good education. It is delightful to realise that

through education human nature will be continually improved, and brought to such a condition as is worthy of the nature of man. This opens out to us the prospect of a happier human race in the future' (8). Needless to say, happiness continues to elude the human race and this text would require a great deal more commentary in another context.

10 A new culture to suit the world

1 Michel Serres, *Statues: The Second Book of Foundations*, trans. Randolph Burks (London: Bloomsbury, 2014).

2 The Manhattan Project was the name of a very secret operation (between 1942 and 1946) run by the United States, Canada and the United Kingdom which allowed the United States to develop the atomic bomb during the Second World War. At its peak there were 130,000 people at work for this project. It was considered a highly prestigious project; among the scientists working on it, twenty received a Nobel Prize (before or after this project).

3 The Seveso disaster was an industrial accident that happened in a small industrial manufacturing plant in the Milan area, Italy. It caused the highest known exposure to 2,3,7,8-tetrachlorodibenzo-p-dioxin (TCDD) in residential populations. It gave rise to numerous scientific studies and standardized industrial safety regulations. The EU industrial safety regulations, for instance, are known as the Seveso II Directive.

4 Serendipity (meaning 'the unforeseen surprise') was coined by Horace Walpole in 1754. In a letter he wrote to his friend Horace Mann, Walpole explained an unexpected discovery he had made about a painting of Bianca Cappello by Giorgio Vasari by reference to a Persian fairy tale, *The Three Princes of Serendip*. The princes, he told his correspondent, were 'always making discoveries, by accidents and sagacity, of things which they were not in quest of'.

5 Part of the Valencia Foundation for the Third Millennium, Serres, in 1999, published the text below which has been implemented by several universities in Paris and which continues to play a role in the ethical and moral duties of science today.

The Scientist's Oath

In the presence of my university masters and before all,
I pledge and I promise to be loyal to the laws of honor and integrity in the exercise of scientific research.

Never shall I falsify the results of my research.

Nor shall I permit questions of religion, nationality, race, social class or political affiliation to interfere with the elaboration or interpretation of my research.

I shall maintain an absolute respect for human life.

Even under threat, I shall never permit my scientific knowledge to be used against the laws of humanity.

Within my power, never shall I allow my knowledge, inventions or their applications to serve violence, destruction, death, increased poverty, enslavement, or inequality.

That I be accorded the rightful esteem of my peers should I remain loyal to these promises and that I be disgraced, should I fail to keep my promises.

See, for instance: http://unesdoc.unesco.org/images/0011/001158/115858e.pdf.

6 Massive Open Online Course.

INDEX